走进辽宁古生物世界

Entering Paleontological World
in Liaoning

走 进
辽宁古生物世界

孙 革 等著

上海科技教育出版社

本书作者

孙　革　胡东宇　周长付　刘玉双　杨　涛
傅仁义　程绍利　杨建杰　张洪钢　孙永山　王丽霞

第一作者简介　First Author

孙　革　辽宁沈阳人，古植物学家，教授、博士生导师，辽宁古生物博物馆馆长，沈阳师范大学古生物学院院长，中国古生物学会副理事长，国家古生物化石专家委员会顾问，美国植物学会通讯会员，美国佛罗里达大学（自然史博物馆）名誉教授，日本城西大学顾问，《世界地质》（*Global Geology*）主编。1968年毕业于长春地质学院（现吉林大学），1985年于中国科学院南京地质古生物研究所获博士学位，1988—1989年于英国大英博物馆（自然史部）完成博士后研究。曾任国际古植物学会（IOP）副主席，中国科学院南京地质古生物研究所研究员、副所长，吉林大学古生物研究中心主任等。从事地质古生物学研究40余年，为全球早期被子植物研究作出突出贡献；1998—2002年率课题组首次发现迄今世界最早的被子植物"辽宁古果"和"中华古果"，确认"古果属"（*Archaefructus*）为水生草本被子植物，提出"被子植物起源的东亚中心"假说，有力地推动了我国及全球被子植物起源及早期演化研究。曾获"教育部自然科学一等奖"、"李四光地质科学奖"及"辽宁省科学技术一等奖"等，2014年获"全国优秀科技工作者"称号。已发表专著9部（包括合作）、译著1部、论文160余篇。

目录/Contents

前言 ·· 1

第一章 走进辽宁古生物世界 ·· 4
  1.1 30亿年来的地质变迁 ·· 4
  1.2 辽宁"十大古生物群" ·· 14

第二章 辽宁古生物博物馆巡礼 ·· 64
  2.1 博物馆设计与展陈 ·· 66
  2.2 五年建设历程 ·· 74
  2.3 十大化石明星 ·· 80
  2.4 闪亮的科研 ·· 91
  2.5 活跃的科普 ·· 97
  2.6 国内外交流 ·· 108
  2.7 人才培养 ·· 119
  2.8 实验室建设 ·· 126
  2.9 博物馆管理 ·· 129

第三章 辽宁化石保护工作 ·· 132
  3.1 辽宁化石保护工作 ·· 132
  3.2 博物馆共建 ·· 136

第四章 辽宁建立古生物学院 ·· 138
  4.1 古生物人才培养的新摇篮 ·· 138
  4.2 学院与博物馆 ·· 143

致谢 ·· 145

Entering Paleontological World in Liaoning

Preface ··· 151
Chapter 1 Entering Paleontological World in Liaoning ··· 154
   1.1 Geological changes in the past three billion years ··· 154
   1.2 Top Ten Fossil Biotas of Liaoning ··· 158
Chapter 2 Introduction to Paleontological Museum of Liaoning ································· 169
   2.1 The PMOL design and exhibition ··· 170
   2.2 Review of PMOL in the past five years ··· 174
   2.3 Ten Fossil Stars ··· 176
   2.4 Fruitful scientific research ··· 180
   2.5 Active scientific popularization ··· 181
   2.6 Communication and cooperation ··· 183
   2.7 Personnel training ·· 185
   2.8 Laboratories and museum management ··· 187
Chapter 3 Fossil Protection in Liaoning ·· 188
   3.1 Fossil protection in Liaoning ·· 188
   3.2 Co-building the PMOL ··· 189
Chapter 4 College of Paleontology built in Liaoning ·· 190
   4.1 New cradle for training young paleontologists ·· 190
   4.2 The college with PMOL ·· 192

参考文献/Reference ·· 193

前　言

我们生活的地球经历了约46亿年的漫长的演变,至今没有"停歇"。距今约2.6亿年起,地球随着大气圈二氧化碳浓度的增加开始向"温室"过渡。而后在距今约300万年的新生代晚期起,又向"冰室"气候转化。亚洲季风的形成和加强使岩石圈陆地沙漠化日趋严重;至今,水圈中水的污染在不断加重;生物圈中生物多样性在日益减少,等等。所有这些地球变化都严峻地摆在人类的面前。更好地认识和改善地球变化所带来的影响,了解46亿年以来地球变化的起因、过程和规律,特别是自38亿年以来地球生物界及其环境的演变,从中找出一些规律或启示,这不仅是地学工作者,特别是古生物工作者的主要任务之一,也是广大科学爱好者在知识上的期盼。

在距今约38亿年,地球上的原始海洋中出现最早的生命——蓝藻和细菌。这些早期生命从单细胞到多细胞、从简单到复杂,不断演化;在距今6亿多年出现最早的后生动物,又在距今4亿多年出现最早的陆地维管植物。从此地球告别了最初的单调和寂寞,逐渐发展为今天生机盎然的世界。但地质历史时期生命如此漫长的演化历程是如何得知的?这些生命演化过程中有哪些精彩的故事?给我们哪些有益的启迪?所有这些都是靠保存在岩石中的生命遗体或遗迹——古生物化石得知的。化石,这些大自然留给人类的宝贵遗产,记录了亿万年来地球上发生的海陆变迁和生命演化的轨迹,也向我们展现了它们所经历的一个个神奇的故事。

辽宁是我国、乃至东亚地区地质历史和古生物化石最古老的地区,其最早的岩石记录距今已38亿年,最早的古生物记录也可追溯到距今25亿年鞍山群的原始藻类化石。因此，走进辽宁的地质古生物世界，可以说是走进了认识我国地质和生命演化史最具代表性的"大门"之一。到辽宁看化石，不仅有助于领会广博的地质古生物知识，而且又能在游弋知识的海洋时产生无尽的遐想并思索地球以及人类的未来。

2006年，辽宁这个"古生物大省"出现了一项古生物学事业上的壮举：辽宁省国土资源厅和沈阳师范大学开始共建我国迄今规模最大的古生物博物馆——辽宁古生物博物馆。经过五年的努力，该馆于2011年5月21日正式完工并对外开放。这全然一新的博物馆以宏伟的建筑气势和丰富的科学内涵，向世人展示被誉为"世界级化石宝库"的辽宁古生物化石，许多化石是"世界级珍宝"。在这里，你可以亲眼见到举世闻名的热河生物群和燕辽生物群的珍稀化石，诸如世界最早的带羽毛恐龙——赫氏近鸟龙，迄今最早带毛发的似哺乳动物——巨齿兽，迄今最早的花化石——辽宁古果和中华古果，迄今最早的会滑翔的蜥蜴——赵氏翔龙，迄今辽宁最大的恐龙——辽宁巨龙，以及近年来新发现的原始鸟类沈师鸟、渤海鸟、盛京鸟以及龟类化石新类群辽龟、小龟等。种类之多、内容之丰富，令人目不暇接。这些珍贵的化石为解决许多涉及全球生命演化的重大理论问题（如鸟类起源、被子植物起源、哺乳动物的早期演化等），作出了重要贡献。近20年来，有关辽宁古生物化石的许多研究成果不仅发表于英国的《自然》(Nature)、美国的《科学》(Science)等世界权威学术杂志，还多次入选我国十大科技新闻（进展）及美国的"百大科学新闻"等。由于这些化石的发现，辽宁被誉为地球上"第一只鸟起飞、第一朵花绽放的地方"。因此，参观和介绍辽宁古生物博物馆已成为我国科学文化，特别是科学普及中一件非常有意义的活动，本书的写作和出版也应运而生。

本书包括四章15节，就内容而言，她是2011年出版的《30亿年来的辽宁古生物》一书的姊妹篇，但本书对辽宁"十大古生物群"的内容做了大量的补充，特别是对燕辽生物群和热河生物群做了更详细的介绍，并对寒武纪—奥陶纪生物群、本溪生物群、阜新生物群和抚顺生物群等做了新的补充，还增加了部分相关科学家的简介。此外，对辽宁古生物博物馆及其展品进行了较详细的介绍，并首次推出辽宁古生物博物馆展品中的"十大化石明星"，特别是对迄今全球最早的"鸟"——世界最早的带羽毛恐龙赫氏近鸟龙，以及迄今全球最早的"花"——辽宁古果，进行了较为详尽的讨论，包括它们的发现对研究全球鸟类起源、被子植物起源的重要意义以及近年来在国际上的最新影响的介绍等。本书第二章集中回顾了辽宁古生物博物馆的建馆历程,特别介绍了该馆的办馆理念、重要科研成果、活跃的科普活动以及广泛开展的国内外交流活动,还包括对参与"政府与高校共建博物馆"有益尝试中的体会;也对辽宁省的化石保护及新建的我国首家古生物学院与博物馆工作的密切配合等做了点赞。

总之,本书是一部较大型的科普作品,又是一部带有较强学术色彩的、有关辽宁古生物的知识读物。作者们凭借数十年在辽宁开展地质古生物工作的学术积累,并结合近年来在辽宁古生物博物馆的工作实践,不仅深入浅出地带读者走进辽宁30亿年来奇妙的古生物世界、领略辽宁古生物的多样性,而且以崭新的科学视角展现了我国地质古生物博物馆事业的新发展及向"国际化"方向办馆的一些新理念。广大读者不仅将从本书中得到地质古生物知识的熏陶,而且可能会在建设博物馆工作思路等方面得到一些有益的启迪。

2016年5月21日,辽宁古生物博物馆迎来了建馆五周年纪念日。本书的出版作为献礼之一。作者们衷心感谢国土资源部、辽宁省国土资源厅、辽宁省化石资源保护管理局和沈阳师范大学多年来予以的全力支持和指导;衷心感谢辽宁古生物博物馆和沈阳师范大学古生物学院全体同事的协助,特别感谢段吉业、郑少林、张立君、王五力及张宜等教授对本书所作的贡献;也由衷感谢彭善池、董枝明、王成源等教授对本书写作的指导与帮助。借此机会,也向国内外诸多专家学者对辽宁古生物博物馆工作所给予的热诚关心和支持表示深深的谢意。

本书的出版得到国土资源部东北亚古生物重点实验室、教育部东北亚生物演化与环境重点实验室、辽宁省古生物演化与古环境变迁重点实验室、沈阳师范大学与中国科学院南京地质古生物所及古脊椎动物与古人类研究所共建的"东北亚古生物协同创新平台"的支持。本书作者还特别感谢上海科技教育出版社王世平总编、伍慧玲及汤世梁编辑,以及卞毓麟教授等在编辑出版工作中给予的鼎力支持。

辽宁古生物博物馆馆长
沈阳师范大学古生物学院院长
2016年8月于沈阳

第一章

走进辽宁古生物世界

1.1 30亿年来的地质变迁

古生物化石是地球地质演化进程中的产物之一,地质时期生命的诞生与演化离不开它们所处的地质变迁背景。辽宁是我国地质历史最古老的地区之一。辽宁陆块是华北陆块(地盾)的东延部分,其形成初期(主要为太古宙)岩石圈薄,活动性较强,热流值高,海底火山喷发不断;经迁西运动(约35亿—31亿年前)[39]后,出现砂岩等沉积岩,显现较稳定的浅海环境,范围不断扩大。辽宁鞍山地区太古宙的石英闪长质和奥长花岗质岩是中国已知最古老的岩石之一,同位素测年值约38.4亿年[39],代表了华北陆块结晶基底的部分时代。这套古老的地层称鞍山群或建平群,多为麻粒岩相的海底喷发镁铁质岩、酸性火山岩和沉积岩夹磁铁石英岩(构成鞍山式铁矿);鞍山地区陈台沟组的锆石年龄约33.5亿年(据张立君面告,2016)。太古宙晚期海水中蓝绿藻、铁细菌以及其他原始藻类的出现和氧气释放是鞍山群大规模铁矿形成的主要原因之一。太古宙末鞍山运动(距今约25亿年)使基底变质和强烈变形,伴有大面积片麻状花岗岩侵入,硅铝质的地壳最终形成,成为陆核的一部分。至元古宙早期结束时,辽宁陆块(陆台)基本固结形成。由于辽宁的初始地壳与华北地区相连,因此统称华北陆台。华北陆台形成这一地质事件奠定了辽宁的地质基础。

太古宙总体为缺氧的还原型大气环境。但新太古代—古元古代由缺氧向含氧过渡,主要由于海水中蓝藻和铁细菌的作用。这些原核生物在海水中逐渐繁盛、释放氧气,使海水中的二价铁离子(Fe^{2+})吸收氧气变成三价铁离子(Fe^{3+})沉淀;这一化学变化促使海水中三氧化二铁(Fe_2O_3)增加,形成大量赤铁矿,乃至进一步形成磁铁矿[46]。辽宁鞍山群新太古代地层及古元古代辽河群下部地层中产出条带状磁铁石英岩、大量铁矿及其相关矿产,是辽宁地质史上的一大重要事件。

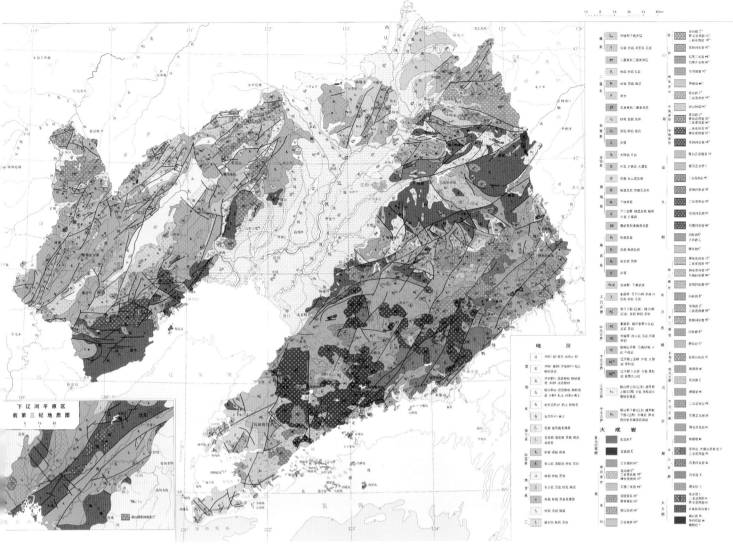

辽宁省地质图(据辽宁省地质矿产局,1999)

辽宁本溪南芬太古宙鞍山群鞍山式铁矿

孙革教授(左1)在南芬介绍鞍山群

李廷栋院士(左)等考察南芬鞍山群

距今约25亿—16亿年,辽宁的元古宙早期地层主要分布于辽南地区,以辽河群为代表。中元古至新元古代早期,华北陆台曾发生大规模开裂,形成多个海槽,在辽宁幅域,主要表现为"燕辽海",沉积厚度上万米,主要分布在辽西,此时的辽南和辽东则大部分为古陆。"燕辽海"形成的沉积分别被称为长城系、蓟县系和青白口系。中元古代地层主要以长城群及蓟县群为代表,距今约18亿—10亿年;新元古代早期(青白口纪)沉积距今约10亿—8亿年。这一时期的古生物化石主要表现为叠层石等[39]。

新元古代中晚期,即南华纪(距今8.0亿—6.85亿年)和伊迪卡拉纪(即原"震旦纪",距今6.85亿—5.42亿年),辽宁南北两侧陆块出现了明显抬升,海水主要集中于辽南和辽东地区,为滨海和海滨湿地。其地层按时代先后,分别被称为永宁组、钓鱼台组、南芬组、桥头组、五行山群和金县群等。辽南大连金州地区已发现水母类、浅海底栖钻孔动物及蠕形动物的遗迹化石等,这些后生动物的时代距今约6.0亿—5.7亿年。此阶段的生物仍比较单调,以菌藻类为主,并形成大量叠层石藻礁。辽南后生动物群的发现对研究辽宁早期生命演化以及与伊迪卡拉型生物群的对比研究等具有重要意义。

寒武纪时期,辽宁幅域曾大部分被海水覆盖;奥陶纪早中期基本继承了寒武纪的海陆格局,但在辽西和辽北有火山活动[39]。受加里东构造运动的影响,中奥陶世末,包括辽宁在内的华北陆

王鸿桢,山东苍山人,我国著名地质古生物学家,中国科学院院士。1939年毕业于西南联大,1947年获英国剑桥大学博士学位。曾任北京大学地质系教授,北京地质学院副院长,武汉地质学院院长,中国地质大学教授,中国地质学会秘书长、理事长等。他为我国地层学、沉积古地理学和古生物学研究与教学作出杰出贡献,是中华人民共和国地层古生物教育事业的开拓者之一、历史大地构造学的奠基人之一。他所编撰的《中国各地质时期古地理图》一直是我国地学界的经典。

王鸿桢(1916—2010)

台地壳强烈抬升,持续近1.4亿年(距今4.6亿—3.2亿年),以致缺失志留纪和泥盆纪沉积。华北陆台自早石炭世晚期起(距今约3.2亿年)才再次遭受古亚洲洋海侵,辽宁大部分地区成为滨海和浅海。但海水时进时退,成为海陆交互相沉积,主要沉积地层为本溪组和太原组(可能为一部分),主要分布于辽

考察本溪牛毛岭石炭纪地层剖面

东和辽南,在本溪牛毛岭和大连复州湾有晚石炭世沉积的典型剖面[17]。在本溪牛毛岭剖面,在中奥陶世马家沟灰岩之上,呈现明显的不整合界面。界面之上为紫色含铁质页岩及G层铝土矿,代表了风化剥蚀沉积的产物。这层紫色含铁质页岩及G层铝土矿在本溪五湖咀煤田的本溪组出露3—6米,在整个华北地区也有广泛分布。本溪组最早由我国地质学家赵亚曾(1925)在本溪牛毛岭剖面建立,著名地质古生物学家李四光、盛金章院士等均在此做过研究工作[39]。

早石炭世晚期,由于辽宁地处滨海和浅海环境,气候温暖潮湿,植被逐渐茂密,在滨海沼泽地区出现森林,形成一些含煤沉积。至晚石炭世,辽宁东南部和辽西的海水已不同程度地退出,华北陆台出现广泛的内陆河湖盆地,形成大批沼泽。由于当时气候炎热潮湿,植被繁茂,森林密布,成为辽宁的"第1次主要成煤期"(晚石炭世—早二叠世)[39]。二叠纪主要含煤地层包括太原组(或一部分)、山西组(主要含煤层),下石盒子组也含少量煤层。整个华北地区晚石炭世—早二叠世含煤植物群被称作"华夏植物群"[19]。

二叠纪晚期,受海西运动影响,辽宁的海水已全部退出,地壳进一步抬升,形成了中国

本溪市政府在牛毛岭剖面为李四光院士(中)、赵亚曾(左)及盛金章院士建立塑像(2007)
李廷栋院士(第二排左9)等出席揭幕仪式。

东部、北部一些零星的小型内陆盆地,主要分布于辽东本溪和辽西北票等地。在辽宁,其代表性地层为以红色沉积为主的郑家组(相当于华北的原石千峰组)的下部。郑家组的上部可能已进入中生代早三叠世。

中生代初期的早三叠世气候干旱,河湖范围缩小,主要为红色或杂色河流相沉积。这一时期的地层在辽西主要以喀左杨树沟等地的红石砬子组为代表,在辽东主要以郑家组上部的红色沉积为代表;此期,整个华北、西北地区均较为相似,此次干旱事件带有全球性质,早三叠世的植被以石松类肋木(*Pleuromeia*)等为特色。进入中三叠世后,其陆相地层及生物群发现较少,但发现于本溪林家崴子的中三叠世林家组生物群独具特色。林家组见于本溪市西南的林家崴子,主要由灰白、黄绿色为主的砂砾岩夹紫色和灰黑色粉砂岩及页岩组成,平行不整合覆于晚二叠世—早三叠世郑家组以之上,富含植物等化石。

到了距今约2.3亿年,辽宁进入晚三叠世,气候总体上转为温暖而潮湿。至侏罗纪早中期,辽宁大地河湖遍布,生物繁盛,森林密布,此间伴有多次强烈的火山活动,形成大量

李四光(1889—1971)

李四光,湖北黄冈人,著名地质古生物学家,中国地质学创始人之一,中国科学院学部委员(院士)。1919年和1927年分别获英国伯明翰大学硕士和博士学位。曾任中央研究院地质研究所所长、中华人民共和国成立后任中国科学院副院长、地质部部长、中国科协主席。他为我国地层古生物研究、第四纪冰川和构造地质学研究以及为大庆、胜利等油田的发现作出了杰出贡献。1926年他与赵亚曾共同创名"本溪系",1927年他对本溪牛毛岭剖面"本溪系"的䗴类开展了详细研究。

赵亚曾(1898—1929)

赵亚曾,著名地质学家。1925年来本溪进行地质调查,在其《南满石炭纪地层之研究》报告中首次命名了"蚂蚁石灰岩"、"小峪石灰岩"和"本溪石灰岩"等层级岩石地层单位,并根据灰岩中含有腕足类 *Spirifer mosquensis*,确定其时代为中石炭世;1926年正式创名"本溪系"。

盛金章(1921—2007)

盛金章,江苏靖江人,著名古生物学家,中国科学院院士。1946年毕业于重庆大学。原中国科学院南京地质古生物研究所研究员。他最早建立中国二叠纪䗴类化石带,1951年在牛毛岭剖面首次命名"牛毛岭石灰岩",1958年他首次发现"小市灰岩"。

晚石炭世本溪植物群景观与植物化石
1. 晚石炭世景观复原；2, 3. 梭鳞木（*Lepidodendron szeianum*）。
（据孙革等，2011）

本溪林家崴子郑家组（P_3–T_1）红色沉积（右）及中三叠世林家组（T_2l，黄线之上）

含煤沉积盆地和含煤地层。早侏罗世含煤地层在辽西称北票组，在辽东称长梁子组[67]。辽西的北票组之下为早侏罗世底部的火山岩地层——兴隆沟组，以安山岩为主，厚200—400米，时代大约为距今191Ma*；在辽东，早侏罗世底部为北庙组，为灰色紫红色凝灰质砂砾岩及灰色安山岩。上述火山岩代表了辽宁在早侏罗世曾出现一次较大规模的火山活动。北票组及长梁子组富含具工业价值的煤层，为辽宁"第2次主要成煤期"。北票组富含植物化石，我国古植物学家米家榕等曾就北票植物群及地层做过大量研究工作[105]。

中侏罗世时期（距今约182—163Ma）的辽宁，火山活动较为剧烈，在初期的较为稳定的河湖相沉积（以海防沟组为代表）之后，在辽西主要以髫髻山组（即兰旗组）为代表，该组

---

* Ma指"百万年"。

在本溪田师傅早侏罗世长梁子组采集化石

地层主要由灰紫、暗灰色安山岩夹凝灰质砂岩及粉砂岩等组成,厚约786米。粉砂岩中含丰富的动、植物化石,形成著名的"燕辽生物群"[39]。

在建昌玲珑塔大西山剖面,胡东宇、徐星等首次发现迄今世界最早的带羽毛恐龙赫氏近鸟龙(*Anchiornis huxelyi*)[74],罗哲西、季强等首次发现迄今最早的真兽类哺乳动物中华侏罗兽(*Juramaia sinensis*)[79]。此外,近年来在这一层位中还发现大量的翼龙、鱼类、叶肢介、介形类、双壳类、腹足类、昆虫及植物等化石。这些燕辽生物群化石的发现为研究鸟类起源和真兽类起源等作出了重要贡献。

度过了晚侏罗世相对较为干旱的时期(即相当于土城子组时期),辽宁进入了中生代真正的"温室"时期[45]。在早白垩世早中期(距今约135—120Ma),辽宁又一次经历了大规模强烈的火山活动,主要以辽西的义县—北票—凌源一带为代表,火山活动形成一批山间盆地,主要沉积物为火山岩夹含火山质的沉积碎屑岩,以义县组(距今约135—122Ma)和九佛堂组(距今约122—120Ma)为代表。此间,河湖遍布,气候温暖(或具有季节性干旱),动、植物十分繁茂,产著名的"热河生物群",包括我国最早发现的带毛恐龙[中华龙鸟(*Sinosauropteryx*)[16]、小盗龙(*Microraptor*)[91]等]、原始鸟类[孔子鸟(*Confuciuornis*)[12]]以及迄今最早的被子植物[辽宁古果(*Archaefructus liaoningensis*)[85]、中华古果(*A. Sinensis*)[86]等]等一批珍贵化石,为我国及全球中生代生物演化研究提供了珍贵的化石依据。

热河生物群的演化包括大北沟期(早期萌发阶段)、义县期(中期辐射演化阶段)及九

米家榕(1930—2013)

米家榕,天津人,古植物学家。1952年毕业于北京大学。原长春地质学院地层古生物教研室主任、教授,天津师范大学教授;曾任中国古植物学会副理事长、中国古生物学会东北地区专业委员会主任等。他为我国古植物学、中生代地层学等研究与教学作出重要贡献。自20世纪50年代起,他在研究辽西北票组植物群及地层中作出重要建树;70年代于吉林通化首次发现我国"北方型"晚三叠世植物群;他的《冀北辽西早、中侏罗世植物古生态学及聚煤环境》及《辽东太子河流域早石炭世植物古生态与古环境》等著作为研究辽宁古生物与地层作出重要贡献。

国内外专家考察大西山剖面(2015)

佛堂期(晚期萎缩消亡阶段)等3个阶段[39]。早期热河生物群属古大兴安岭—额尔古纳河生物地理区系,总体上处于热河生物群的萌发阶段,以大北沟期尼斯托叶肢介—三尾拟蜉蝣—滦平介(Nestoria-Ephemeropsis trisetalis-Luanpingella)生物组合为代表,以叶肢介和介形类最为丰富,另有少量的鱼类、双壳类、腹足类和鲎虫等;生物群的分布范围不大,主要为沿冀北—大兴安岭南北向狭长地带分布。中期阶段以义县期的生物化石为代表,几乎包括了目前已发现的热河生物群的所有门类,是热河生物群发展的高峰期和快速辐射期;以孔子鸟—中华龙鸟—狼鳍鱼(Confuciusornis-Sinosauropteryx-Lycoptera)生物组合为代表,分布范围以燕辽地区为中心,扩展至以中国北方为主体的东亚地区,包括蒙古、俄罗斯外贝加尔等地。晚期阶段以九佛堂期的华夏鸟—延吉叶肢介—格氏湖女星介(Cathayornis-Yanjiestheria-Limnocypridea grammi)生物组合为代表,化石数量和类别也还比较多,分布范围比中期阶段更大,但总体上显示热河生物群已从高峰期转向萎缩和消亡。

早白垩世晚期火山活动形成的山间盆地,由于气候转为湿润温暖,森林茂密生长,形成辽宁"第3次主要成煤期",以辽宁北部的阜新、铁法等地的煤田为代表,时代约为距今

建昌玲珑塔大西山侏罗纪剖面(箭头示产赫氏近鸟龙的化石点)

热河生物群演化阶段及地理分布示意图（据陈丕基，1988，修改）

北票四合屯义县组的火山岩侵入体（箭头示）

110Ma[39]。

古近纪之初（距今约66—50Ma），受太平洋板块和欧亚板块碰撞等影响，中国东部大陆发生了一系列的断陷与隆升：下辽河一带形成大型的裂谷盆地，抚顺等地形成地堑盆地。古近纪之初的火山活动以抚顺老虎台组的基性火山岩和栗子沟组的火山凝灰岩最具代表性；火山喷发间歇期，由于气候温暖潮湿，抚顺盆地河的湖泊和沼泽遍布，森林茂密，形成了老虎台组的B煤组和栗子沟组的A煤组[64]。

进入始新世（距今约50—36Ma）、特别是始新世中期，受全球性升温影响，抚顺生物群的发展达到鼎盛期；此间，辽宁大地的气候炎热潮湿，植物茂盛，森林密布，在抚顺盆地出现沙巴榈（*Sabalites*）、苏铁（*Cycas*）等热带—亚热带植物化石[56,82]，形成了巨厚的煤及油页岩，是辽宁的"第4次主要成煤期"[39]。特别需要指出的是，在计军屯组沉积时期（距今约47.5Ma[80]），抚顺盆地气温和湿度仍较高，由于盆地稳定沉降，积水加深成相对封闭的湖泊，湖水水动力条件较弱，湖盆中低等生物繁盛，其遗体及黏土物质在还原条件下，经化学、物理和地质作用而形成巨厚的油页岩（厚度可达48—190米）。至始新世晚期（以耿家街组为代表，距今约38—36Ma），由于气温逐渐降低，盆地基底有所抬升，湖盆逐步向淤塞方向发展，抚顺生物群逐步进入萎缩期。

新近纪（距今2300万—181万年）结束后，进入第四纪（距今181万—约1万年），辽宁基本上继

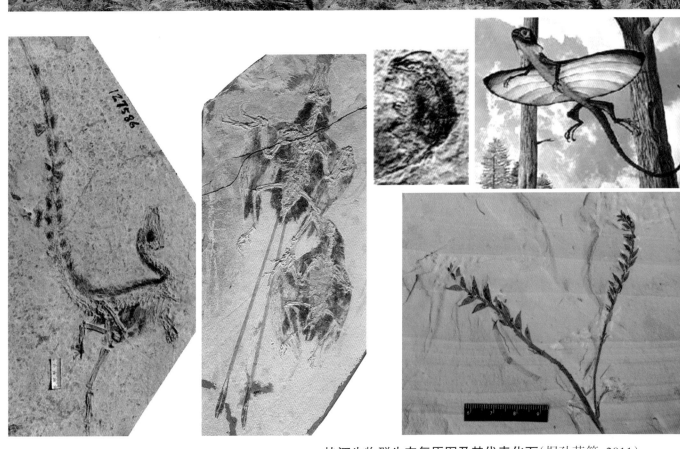

热河生物群生态复原图及其代表化石(据孙革等,2011)

承了前期的古地理格局,下辽河地区持续下降,东、西山地抬升;之后,经历了更新世第1次冰期、渤海第1次海侵、地壳抬升及规模较小的山区冰川及下辽河第2次海侵等多次事件。在全新世(距今约1万年)冰期结束后,最终形成现今的地理格局。辽宁的第四系以松散堆积的砂砾层、黄土层为主,局部地区尚见有火山岩。自更新世以来,由于受新构造运动等诸多因素影响,既有冲洪积物,又有冰碛、冰水沉积物以及洞穴堆积等,古人类得到发展。在本溪庙后山、营口金牛山等地已发现古人类化石,包括直立人阶段的本溪"庙后山人"(距今约50万—45万年)及营口"金牛山人"(距今约28万年)等;伴生的哺乳动物群主要以猛犸象—披毛犀动物群为代表。前不久，在大连复州湾金远洞又新发现可能较之本溪庙后山时期更早的哺乳动物群，其古人类化石也在寻找之中。到距今10万—1万年的晚期智人阶段，辽宁的古人类得到进一步发展。随着末次冰期来临，辽宁的古人类可能从亚洲东北部跨过白令海峡陆桥，踏上北美大陆。我国东北、朝鲜半岛及日本诸岛的人类，可能都是在晚期直立人阶段的庙后山人或较之更早的古人类基础上发展起来的[15,20,39]。

1.2 辽宁"十大古生物群"

辽宁是我国古生物化石历史记录最早的省份，也是我国古生物化石最丰富的省份之一。迄今在辽宁发现的化石近30个门类、1万余种，主要包括菌藻类、水母类、三叶虫、笔石、古杯、珊瑚、海绵、苔藓虫、腕足类、头足类、籢类、牙形刺、海百合、腹足类等海生无脊椎动物，两栖类、爬行类、恐龙、鸟类、哺乳动物、鱼类等陆生脊椎动物，双壳类、腹足类、叶肢介、介形类、虾类、昆虫、蜘蛛等陆生无脊椎动物，以及植物（包括孢粉）等，其化石发现总量居我国之首[39]。

上述丰富的古生物化石揭示了辽宁在近30亿年的漫长地质时期中，经历了太古宙鞍山群原核生物阶段、元古宙真核生物及多细胞后生生物孕育的时代（包括伊迪卡拉生物阶段），以及寒武纪—奥陶纪生物群的繁荣和晚古生代本溪生物群的接续等海生生物稳定发展阶段，也经历了自中生代以来陆生生物群（主要以林家生物群、羊草沟生物群、燕辽生物群以及热河生物群等为代表）的繁荣发展，阜新生物群、抚顺生物群等晚中生代—新生代生物群的后续演化，以及第四纪哺乳动物与古人类的繁荣。化石也反映了辽宁自生命出现以来，一直是较为适于生物生存和繁衍的地方。因此，辽宁的古生物演化与发展似堪称全球生命演化与发展在东亚地区的一个典型代表。

纵观辽宁古生物化石的科学意义和产出情况之后，孙革等（2011）首次提出30亿年来辽宁最具特色的"十大古生物群"，分别是：（1）太古宙鞍山群早期生命（距今约30亿—25亿年）；（2）早古生代寒武纪—奥陶纪海生生物群（距今约5.3亿—4.6亿年）；（3）晚古生代本溪生

阜新海州露天矿早白垩世阜新组含煤地层

抚顺西露天矿古近纪地层(左)及中外学者考察(2006)

物群(距今约3.3亿—3.1亿年);(4)中三叠世林家生物群(距今约2.4亿年);(5)晚三叠世羊草沟生物群(距今约2.1亿年);(6)侏罗纪燕辽生物群(距今约1.8亿—1.5亿年);(7)早白垩世热河生物群(距今约1.4亿—1.2亿年);(8)早白垩世阜新生物群(距今约1.1亿年);(9)古近纪抚顺生物群(距今约0.5亿年);(10)第四纪古人类化石群(距今约50万—1万年)[39]。

在这十大古生物群中,举世瞩目的中生代"热河生物群"、"燕辽生物群"、太古宙"鞍山群早期生命"和"第四纪古人类化石群"为上述十大古生物群中的主要"亮点"。

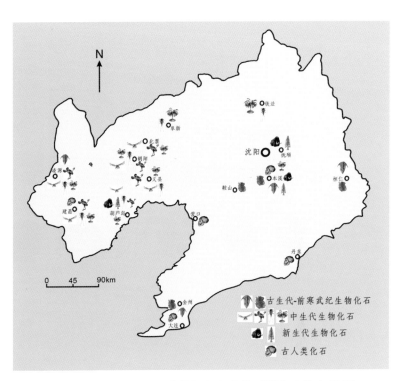

辽宁"十大古生物群"地理分布简图

1.2.1 太古宙鞍山群早期生命

鞍山群是我国最早的地质历史记录之一,主要分布于辽宁鞍山弓长岭至本溪南芬一带,距今约30亿—25亿年;这里的变质岩岩石中已发现原始藻类和蓝藻等化石,这是我国迄今最早的生物化石记录,在全球早期生命演化研究中也占有重要地位[39,49]。

鞍山群早期生命中,目前主要发现是原始藻类化石,他们多分散保存。除原始藻类外,可能还含有蓝藻等原核生物化石。蓝藻或原始藻类是简单的单细胞生物,它们没有(或至少没发现)细胞核,它们大多数的细胞壁外面有胶质衣(膜),可进行光合作用,释放氧气。在海水中的蓝藻和原始藻类等共同作用下,这些原核生物在海水中逐渐繁盛、释放氧气,使海水中的二价铁(Fe^{2+})吸收氧气变成三价铁(Fe^{3+})沉淀(Fe_2O_3),形成大量赤铁矿,乃至进一步形成磁铁矿[46]。

太古宙鞍山群早期生命之后,地球生命逐渐演化为元古宙时期(距今约25亿—5.42亿年)的真核生物及多细胞的后生生物。在辽宁主要表现为见于辽南、辽东地区的叠层石(stromatolite)化石,它们是由蓝藻等原核生物所建造的有机沉积结构,是在蓝藻等生命活动所引起的周期性矿物沉淀、沉积物的捕获和胶结作用下,形成的叠层状生物沉积构造,在辽南及辽西地区都有广泛分布。

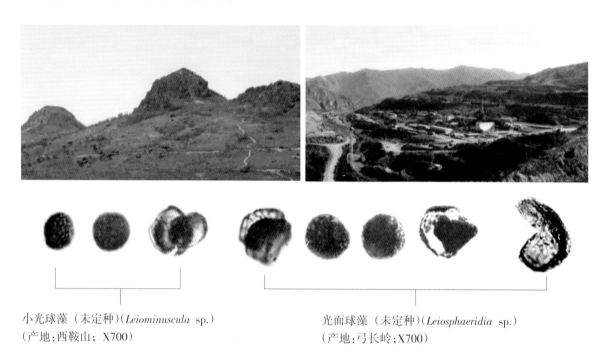

小光球藻(未定种)(*Leiominuscula* sp.)
(产地:西鞍山;X700)

光面球藻(未定种)(*Leiosphaeridia* sp.)
(产地:弓长岭;X700)

西鞍山(左上)及辽阳弓长岭(右上)铁矿鞍山群的原始藻类化石(据尹磊明,1977)

尹磊明,湖南湘潭人,微体古生物学家,中国科学院南京地质古生物研究所研究员,1963年毕业于南京大学。专长于疑源类化石研究,他为我国前寒武纪疑源类及早期生命研究作出突出贡献。1977年他首次在辽宁鞍山群发现原始藻类化石,有力地推动了我国太古宙地层及早期生命研究;2007年在陡山沱组首次发现最早的动物休眠卵化石,并在五台群变质岩中发现疑源类化石。他的专著《中国疑源类化石》是研究我国疑源类及早期生命的经典之作。

尹磊明(1940—)

1.2.2 寒武纪—奥陶纪生物群

在经历了约7.2亿年前发生的南坨大冰期之后，我国的早期生命演化曾经历了蓝田生物群[92]、瓮安生物群[76]、庙河生物群[4]及高家山生物群[14]等早期后生生物演化阶段。至距今5.42亿年进入了寒武纪初期海生生物"大爆发"[3,13]；此后，又经历了奥陶纪海洋生物多样性"大辐射"事件，海洋中的腕足类、海百合及四射珊瑚等动物兴起。寒武纪始于距今5.42亿年，结束于距今4.88亿年。寒武纪是带壳海生无脊椎动物开始繁盛的时代，生物从无壳到有壳，进化中出现重大飞跃。寒武纪以三叶虫最为繁盛，数量上约占整个生物群的60%。

在距今约5.4亿—4.6亿年的寒武纪—奥陶纪时期，辽宁幅域曾是一片大海，海生生物群十分发育。辽宁的寒武纪地层，迄今尚未发现相当于滇东地区晋宁期、梅树村期及南皋期的最早期沉积，其余均有较完整的出露。

辽宁的寒武纪化石主要出露在辽东的太子河流域、辽西凌源—建昌，以及辽南复州湾等3个地区，主要以三叶虫为代表，还包括腕足类、腹足类、头足类、古介形类、古杯类、软舌螺、海百合、牙形刺、笔石、叠层石等，目前已发现120余属400余种，代表了较为稳定的海相浅水沉积环境*。

寒武纪三叶虫（左）与奥陶纪头足类（右）繁盛的海底世界

* 段吉业. 2011. 辽宁古生代动物化石（未刊）

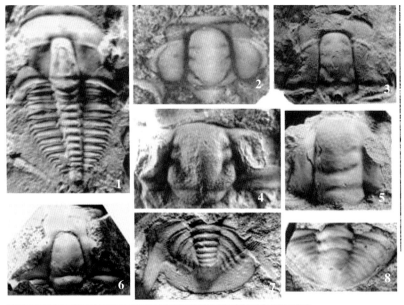

辽宁的寒武纪三叶虫化石
1. 弯曲原波曼虫（*Probowmania curta*）（复县，毛庄组）；2,7. 燕山小蒿里山虫（*Kaolishanella yanshanensis*）（杨家杖子，长山组）；3. 标准定远虫（*Tingyuania typica*）（复县，馒头组）；4. 喀氏拟鲍格朗虫（*Bergeronites ketteleri*）（南票，崮山组）；5,8. 瘤疹褶盾虫（*Ptychaspis pustulosa*）（凌源，凤山组）；6. 刺山东盾壳虫（*Shantungaspis aclis*）（本溪，毛庄组）。（据段吉业，2011*）

本溪寒武纪徐庄组

　　三叶虫是迄今已知地球上出现最早的带壳节肢动物，在寒武纪、奥陶纪最为繁盛，也是划分上述两个纪地层的重要化石依据。辽宁的寒武纪—奥陶纪三叶虫研究已有100余年历史，中华人民共和国成立后，主要有朱兆玲[70]、卢衍豪等[25]、郭鸿俊等[6,7]、段吉业等[5,6]、南润善及常绍泉[29]等研究为代表。近年来，彭善池将寒武系划分为4统10个阶，首建了古丈阶、排碧阶和江山阶底界等3个"金钉子"，有力地推动了国际寒武纪地层划分对比研究[103,104]。辽宁寒武纪与奥陶纪三叶虫群面貌特征有重要更替，主要表现在种类和生物组合方面均不相同。

---

* 段吉业. 2011. 辽宁古生代动物化石（未刊）

除三叶虫外,辽宁寒武纪还广泛产出腕足类化石,它们多为较低等的类群,如舌形贝(*Lingula*)等。头足类化石在辽宁寒武纪晚期也多有发现,头足类壳体主要有直锥形和旋卷壳两大类,辽宁最常见的为小型的爱丽斯木角石(*Ellesmeroceras*)。古杯类动物为典型的净生环境的底栖生物,包括有单体或群体,其内、外壁生有多孔,辽宁复州湾地区寒武纪早期碱厂组已发现古杯类化石,并形成生物礁。此外,辽宁寒武纪还发现有海绵化石,主要为开腔骨针(*Chancelloria*);寒武纪早期还发现棘皮动物灯塔海百合(*Dengtacrinus*)等。笔石类在寒武纪晚期在辽宁已有发现,主要为树笔石类(graptodendroids),本溪田师傅及辽阳均发现有网格笔石(*Dictyonema*)。牙形刺化石也较丰富,主要发现于太子河流域及复州湾地区的寒武纪晚期地层。此外,在辽宁各地的寒武纪地层均发现蓝藻类沉积形成的叠层石。

奥陶纪距今约4.88亿—4.43亿年,是地质史上大陆区遭受广泛海侵的时代,生物界海生无脊椎动物占优势,生物分异性大大增强。中、晚奥陶世也是火山活动、构造运动和冰川活动发育的时代。重要的门类主要为笔石、头足类、腕足类及牙形刺,其次为三叶虫等。

卢衍豪(1913—2000)

卢衍豪,福建永定人,国际著名古生物学家,中国三叶虫研究之父,中国科学院院士。1937年毕业于北京大学。原中国古生物学会理事长、中国科学院南京地质古生物研究所研究员、副所长,国际地层委员会寒武系分会、奥陶系分会选举委员。他最早建立我国寒武系划分标准,确立了我国寒武系为10阶32个化石带;20世纪50年代重新划分东北南部的寒武纪奥陶纪地层;他首创"生物—环境控制论"学说,荣获国家自然科学奖二等奖等。

段吉业(1936—)

段吉业,辽宁沈阳人,古生物学家,吉林大学、沈阳师范大学教授,1961年研究生毕业于长春地质学院。专长于三叶虫研究,为东北地区早古生代三叶虫及地层研究作出重要贡献。他的《华北板块东部早古生代动物群、沉积相及地层多重划分》等著作对辽宁及华北地区早古生代地层及古生物研究作了系统总结。

彭善池(1944—)

彭善池,湖北荆州人,著名古生物学与地层学家。1968年毕业于北京地质学院,1985年获得博士学位。现任国际地层委员会副主席,中国科学院南京地质古生物研究所研究员。他首次提出将寒武系划分为4统10个阶,首建古丈阶、排碧阶和江山阶底界等3个"金钉子"及芙蓉统等,为我国及全球寒武系研究作出杰出贡献。

辽宁的奥陶系仅见其下统(冶里组、亮甲山组,主要为厚层白云质灰岩、白云岩等)和中统(马家沟组,主要为中厚层花纹状灰岩),缺失上统。由于奥陶系大都由巨厚的石灰岩组成,在地貌上常形成大小各异的溶洞,以本溪的"水洞"和桓仁的"望天洞"最为著名。辽宁奥陶纪化石以笔石、头足类及牙形刺为主,其次为三叶虫、腹足类、腕足类、海绵类等。

笔石动物是一类已经绝灭的群体海生动物,个体很小,一般仅长1—2毫米,但个别笔石体也有长达70毫米的。笔石部分营底栖固着生活,有固定的茎、根等构造;也有一部分营漂浮生活,有线管(丝状体),以此附着于浮胞或挂在漂浮的物体上。笔石动物演化迅速,化石分布广,是一类很好的标准化石。辽宁奥陶纪以大量的树形笔石类(gendrograptids)和少许正笔石式的树笔石类为特点。树形笔石按生活方式分底栖和漂浮两类:底栖类有无羽笔石(*Callograptus*)和树笔石(*Dendrograptus*),笔石体以茎连在一个基盘上,彼此独立,向上生长;漂浮的以网格笔石为代表,笔石体互相有连接管相连,形成网状。在辽

辽宁寒武纪古杯化石(1,2)及灯塔海百合化石(3)(据段吉业,2011)

本溪水洞(左)及桓仁望天洞(右)的奥陶纪灰岩溶洞

辽宁奥陶纪动物化石
1. 无羽笔石；2. 树笔石；3. 阿门角石；4. 多泡角石(*Polydesmia*)；5. 链角石(*Ormoceras*)；
1、2产自太子河流域，O_1；3—5产自太子河流域，O_2。(据段吉业，2011)

宁，太子河流域已发现早奥陶世无羽笔石和树笔石，本溪田师傅发现早奥陶世反称笔石(*Anisograptus*)等。

头足类是软体动物门中发育最完善、最高级的一个纲，全为海生的肉食性动物，善于在水底爬行或水中游泳。身体两侧对称，头部两侧具发达的眼，中央有口；腕的一部分环列于口的周围，用于捕食，另一部分则靠近头部的腹侧，构成排水漏斗，也是特有的运动器官。辽宁奥陶纪头足类十分繁盛，太子河地区奥陶系产大量鹦鹉螺化石，如爱丽斯木角石、舒曼角石(*Shumandoceras*)及阿门角石(*Armenoceras*)等。

辽宁奥陶纪腕足动物报道不多，冶里组、亮甲山组以正形贝目(Orthida)为主；上下马家沟组以扭月贝目(Strophomenida)为主。辽宁奥陶纪三叶虫相对寒武纪已大大减少，但这一时期的三叶虫尾甲增大，提高了游泳速度，同时头尾嵌合使整个身体卷曲成球形或盘状，可迅速跌落或潜伏海底，以防侵害。辽宁奥陶纪牙形刺十分丰富，属北方型北美牙形刺生物地理区，已建立了组合序列。在本溪火连寨，奥陶纪牙形石的研究较详细，在生物地层对比研究中发挥了重要作用[1]。

1.2.3 晚古生代本溪生物群

辽宁和华北陆块广大地区从晚奥陶世至早石炭世(距今4.58亿—3.3亿年)基本上同步抬升，缺失了近1亿多年的沉积和生物化石记录。由于晚石炭世整个华北陆块的下降和升降变动，辽宁石炭纪晚期曾是一片浅海，但时有陆地出现，沉积属于海陆交互相。由于当时的气候温暖潮湿，广布的滨海洼地及其周边陆生植物繁盛，真蕨类、种子蕨开始发展，裸子植物苛达树等高大乔木繁盛，成为造煤的重要原料来源之一。这一时期是地质历史中最早的世界性成煤时期,在辽宁也形成了具有一定规模的煤层,被称为辽宁的"第1次主要成煤期"。这一时期的海域范围向东可达朝鲜半岛,向西可达我国山西、宁夏及甘肃等地,基本上均为海陆交互相。至二叠纪早期,辽宁大部分已被抬升为陆地。

长期以来,辽宁一直被认为缺失早石炭世沉积。但米家榕等(1990)在本溪地区本溪组下部G层铝土矿顶板黑色页岩中发现以本溪亚鳞木(*Sublepidodendron*)为代表的植物化石7属12种,认为其时代为早石炭世晚期(维宪期—纳缪尔A期)[27];范国清等据此建立了"高台子组"(厚10—20米);孙革等(2011)认为,该沉积也可以视为本溪组的下部沉积[39]。

辽宁石炭纪本溪组地层厚约150米,海陆交互相沉积中夹5—6层海相灰岩及几层可采煤层,海相层中产有丰富的海相生物化石及大量陆生植物化石。本溪组的典型剖面位于本溪西北郊的牛毛岭,以砂岩为主的地层中夹有(自下而上)"下蚂蚁灰岩"、"上蚂蚁灰岩"、"小峪灰岩"、"本溪灰岩"及"牛毛岭灰岩"等5层以上的海相灰岩地层,化石丰富,盛产䗴、珊瑚、腕足及牙形刺等海相化石,其下部产植物化石,成为我国北方研究石炭纪生物群及地层的理想地区之一。我国著名地质古生物学家赵亚曾、李四光、盛金章等都曾在这里开展研究工作。2007年,本溪市国土资源局在牛毛岭剖面为三位科学家建汉白玉塑像以作纪念。

本溪市国土资源局在牛毛岭建塑像纪念李四光等科学家(2007)

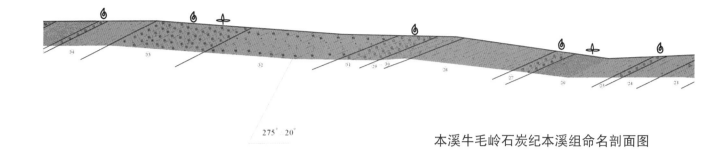

本溪牛毛岭石炭纪本溪组命名剖面图

本溪组最早由赵亚曾(1925)命名,他根据本溪组灰岩中含有腕足类莫斯科石燕(*Spirifer mosquensis*)等化石,将其时代定为中石炭世。盛金章(1951—1958)命名"牛毛岭石灰岩",并将太子河流域本溪组的䗴化石划分为2带5亚带,䗴化石包括著名的纺锤䗴(*Fusulina*)及假史塔夫䗴(*Pseudostaffella*)等[39]。此后,长春地质学院的专家也做过多年大量研究工作:刘发(1987)描述了采自田师傅本溪组下部的腕足类12属18种,包括分喙石燕(*Choristites*)及网格长身贝(*Dictyoclostus*)等[23];林英锡等(1992)系统研究了太子河流域本溪组的皱纹珊瑚化石,共描述珊瑚化石24属43种和3个亚种,并建立2个珊瑚化石组合[22];米家榕等研究了本溪组下部的植物化石等。2005—2007年,孙革、王成源、吴水忠、郎嘉彬等对牛毛岭本溪组建组剖面开展了新一轮研究,根据新发现的牙形刺等化石,确认本溪组上部为国际标准的莫斯科阶的中上部,时代距今约3.1亿年,并认为本溪组的下部可能包括早石炭世的沉积[17, 39]。

牙形刺(Conodonts)是一类已经绝灭的海生动物的骨骼或器官形成的微小化石,外形很像某些鱼类的牙齿或环节动物的颚器,其分类位置至今仍未确定。牙形刺个体微小,一般长0.3—2毫米,主要由薄片状的磷酸钙组成。它们

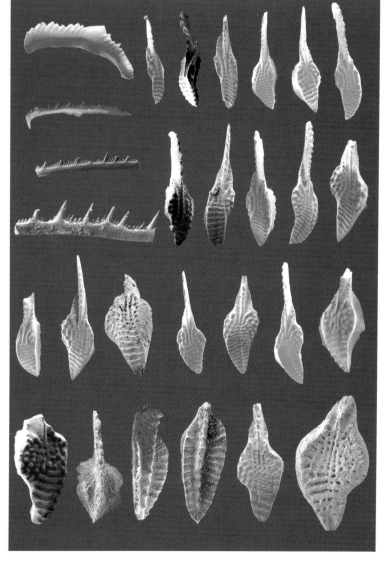

本溪组牙形刺化石

牙形刺化石研究专家
1. (左起)安太庠,王成源,丁惠(1984);
2. 郎嘉彬。

本溪牛毛岭剖面本溪组化石及部分研究专家
右起:俞建章院士,林英铴,武世忠,黄祖熙。

数量多、演化迅速,广布于世界各地,是重要的主导化石门类和地层划分和对比的依据。本溪组牙形刺化石多年来经我国安太庠、王成源、丁惠等多位专家研究,近年来经郎嘉彬与王成源研究,已发现莫斯科阶标准化石泊道斯克异颚刺(*Idiognathodus podolskensis*)、朗德新颚刺(*Neognathodus roundyi*),以及娇柔异颚刺—泊道斯克异颚刺(*Idiognathodus delicates-I. podolskensis*)组合带化石,时代距今约3.1亿年,为本溪组的地层划分、相关太原组时代确定以及提高与国际石炭纪地层对比的精度等作出贡献[17]。

二叠纪早期,辽东的本溪及辽西的南票等地区古生物化石也较为丰富,主要产于太原组(上部)及山西组等陆相地层,其植物化石以早期华夏植物群组成为特色,形成资源丰富的工业煤层。

鉴于辽宁本溪等地区晚石炭世生物群、早二叠世生物群与地层的密切联系,孙革等(2011)建议,晚古生代本溪生物群涵义主要是指出露于辽宁的晚古生代石炭纪生物群(以

俞建章,安徽和县人,国际著名古生物学家,中国科学院学部委员(院士);曾任中央研究院地质研究所研究员、中科院南京地质古生物研究所副所长、长春地质学院副院长。1924年毕业于北京大学,1935年于英国布里斯托尔大学获博士学位,1941年任重庆大学地质系系主任,1947年当选为中国地质学会理事长。他首建中国下石炭统的4个珊瑚带,是我国晚古生代地层及四射珊瑚研究奠基人。

俞建章(1899—1980)

  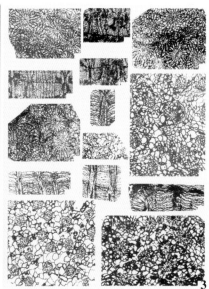

本溪牛毛岭剖面本溪组化石
1. 鋋；2. 腕足类；3. 珊瑚。

本溪地区的化石为主要代表),但也包括辽东和辽西的部分二叠纪早期的生物群[39]。

1.2.4　中三叠世林家生物群

进入中生代(距今约252—66Ma),辽宁已完全变成陆地。中生代包括三叠纪、侏罗纪和白垩纪等三个纪,延续了1亿8千多万年,是地球生物界大变革和大发展时期:以恐龙为代表的爬行类繁盛于生物界,带毛恐龙的演化导致鸟类的出现,翼龙类和原始鸟类占领天空,哺乳动物和被子植物出现并揭开了起源和早期演化的序幕;地球上的生物占领了海、陆、空三维空间,呈现出空前繁荣的景象。在辽宁,由于中生代广泛发育陆相或大陆火山喷发相沉积类型,加之气候温暖,陆生动物和植物、特别是淡水生物门类空前发展,辽宁的生物界呈现了许多奇特的演化。

距今约2.4亿年的中三叠世林家生物群是一个独具特色的生物群,产于本溪林家崴子中三叠世林家组,产地位于本溪市西南、太子河北岸的林家崴子村。林家组生物群的植物化石已发现30余属40余种,主要以束脉蕨—本溪羊齿(*Symopteris-Benxipteris*)组合为代表。伴生化石有孢粉化石具肋双囊粉(*Chordasporites*)、层环孢(*Densoisporites*)、三角孢(*Deltoidospora*)、单沟粉(*Monosulciites*)、罗汉松粉(*Podocarpidites*)、双囊松粉(*Pinuspollenites*)、克拉梭粉(*Classopollis*)等,昆虫化石索德蜰(*Sogdoblatta*)、鱼类化石褶鳞鱼类(*Ptycholepidei*)和古鳕鱼类(*Palaeoniscoidei*),以及轮藻如星轮藻(*Stellatochara*)等化石。时代被认为属中三叠世早期[57,58]。

近年来,张宜、郑少林等对林家组植物群开展了接续研究,新发现较多的、以往常见于晚古生代的分子,如瓣轮叶(Lobatannularia)、东方栉羊齿(Pecopteris orientalis)、舌羊齿(Glossopteris),甚至可能还有大羽羊齿类(Gigantopterids)出现;当然,也发现一些常见于中生代早期的分子,如新芦木(Neocalamites)、丁菲羊齿(Thinnfeldia)及舌叶(?)(Glossophyllum?)等(据张宜,2015*)。

对于林家组植物大化石的时代特征,张武等认为,新建的本溪羊齿(Benxiopteris)虽在营养叶形态上与安加拉二叠纪苏伯羊齿(Supaia)和冈瓦那中三叠世二叉羊齿属(Dicroidium)都有些相似,但生殖器官与后两属明显不同;该植物群中的瓣轮叶(Lobatannularia)、

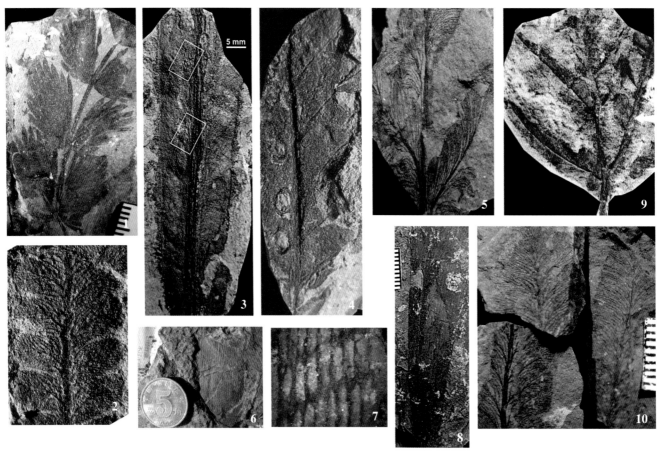

林家植物群化石

1.瓣轮叶;2.东方栉羊齿;3.舌羊齿;4.大羽羊齿(?);5.丁菲羊齿;6,7.新芦木;8.舌叶(?);9.本溪羊齿;10.束脉蕨。(据张宜,2015*)

---

* Zhang Y. 2015. Diversity of the fossil plants from the Middle Triassic of eastern Liaoning, China and its paleoclimatic implication. MTE-12, Shenyang (oral presentation)

本溪林家生物群化石产地(左)及主要研究者(右图左起:张武,郑少林,张宜)

虽最早出现在我国华夏植物群的晚二叠世,但该属在俄罗斯科尔文昌早三叠世植物群、哈萨克斯坦和日本的晚三叠世植物群中也有记载,应被视为古老的孑遗分子[58]。

总之,本溪林家生物群是一个面貌奇特的生物群。就地层层序而言,林家组这套灰白及黄灰色沉积平行不整合

本溪林家生物群化石产地

覆盖于以红色为主的晚二叠世—早三叠世郑家组之上,将其时代考虑为中三叠世早期似有一定道理。但就生物化石(特别是植物群)而言,似不能排除早三叠世或早三叠世晚期的可能。林家组生物群的时代可能还有待进一步研究。

1.2.5 晚三叠世羊草沟生物群

辽宁的晚三叠世地层分布较零星,多以河湖相的砂砾岩及粉砂岩为主,偶夹页岩、薄煤层或煤线。周惠琴(1981)在研究北票羊草沟所采的植物化石时,恢复使用了"羊草沟组"一名[69]。羊草沟组下部地层所产的化石确属晚三叠世,但该组上部的黄绿色砂岩等经对叶肢介化石等的研究,认为时代为早侏罗世。因此,本书所指的晚三叠世生物群是指产

晚三叠世羊草沟生物群产地羊草沟村及含产化石地层

于北票羊草沟组下部的生物群,包括凌源老虎沟及朝阳石门沟等地所产的晚三叠世生物化石。晚三叠世羊草沟植物群除周惠琴[69]研究外,先后又有张武等[55,59]、米家榕等[28]等研究,近年来,孙春林等对羊草沟植物群的研究又取得许多新进展[35,83]。

据郑少林(2011*)研究,辽宁晚三叠世羊草沟植物群共约36属71种植物组成,包括周惠琴(1981)首次报道的北票羊草沟植物化石20属28种,张武报道凌源老虎沟晚三叠世植物11属20余种[55],张武、郑少林研究的朝阳石门沟和北票东坤头营子的24属33种等[59,60]。郑少林认为,该植物群蕨类植物在组合中居首位(11属29种,约占41%),双扇蕨科较为繁盛。松柏类也很丰富(5属12种,约占16.9%),以准苏铁果—苏铁杉(Cycadocarpidium-Podozamites)为特征。银杏类(4属9种,约占13%)属于大的乔木或灌木,叶子季节脱落,反映温暖、潮湿而富于季节变化,陕西舌叶(Glossophyllum shensiensis)被视为重要分子。本内苏铁类(5属7种,约占10%)占一定比例,属喜热耐旱植物。种子蕨(4属4种,约占6%)虽比例不大,但具重要意义,以鳞羊齿(Lepidopteris)和革叶(Scytophyllum)最为重要。综观辽西晚三叠世植物群各大类群的主要成分和性质,可推断它们生长在北半球亚热带至暖温带的气候区内,大体可归入晚三叠世的中国北方植物地理区[37,38]。

晚三叠世羊草沟生物群的孢粉组合经曲立范、蒲荣干及吴洪章[30,31]等研究,共发现54属110种,以本内苏铁类、苏铁类、银杏类及双扇蕨科较丰富,认为反映温暖潮湿的亚热带—暖温带的气候。晚三叠世羊草沟生物群的双壳类、叶肢介及介形类化石也很丰富,主要产于北票羊草沟和凌源老虎沟两个产地。双壳类化石已见至少5属12种,主要以长陕西蚌—宁夏珠蚌(Shaanxiconcha longa-Unio ningxiaensis)组合为代表[57]。

---

* 郑少林. 2011. 辽宁中生代生物群(未刊)

晚三叠世羊草沟植物群化石

1. 新芦木；2. 似木贼（*Equisetites*）；3，7. 蕉羽叶（*Ctenis*）；4. 格子蕨；5. 拟轮叶（？）；
6. 北票叶；8. 准苏铁果；9. 网叶蕨。（主要据孙革等，2011）

近年来，经孙革等（2011）研究认为，晚三叠世羊草沟植物群主要以拟轮叶（？）（? *Annulariopsis*）、新芦木、网叶蕨（*Dictyophyllum*）、格子蕨（*Clathropteris*）、北票叶（*Beipiaophyllum*）[73]、银杏（*Ginkgo*）、苏铁杉（*Podozamites*）及其生殖器官准苏铁果（*Cycadocarpidium*）等为主要代表。双扇蕨科（如网叶蕨、格子蕨等）及大量苏铁类等的出现，代表了我国及东亚地区晚三叠世南方植物群特点；而大量银杏类（如 *Ginkgo*，*Baiera* 等）的出现代表了我国晚三叠世北方植物群的特点。因此，羊草沟植物群总体上似显示中国南、北方晚三叠世植物

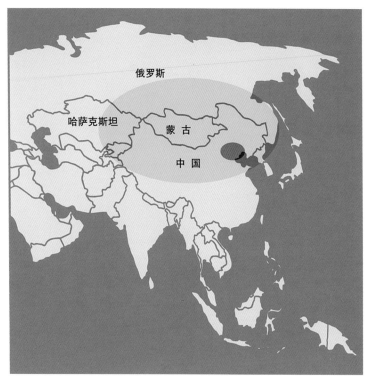

燕辽生物群古地理分布简图(据张宜,2015)

群混生的特征,显示了温暖潮湿的气候。晚三叠世羊草沟植物化石通常形体大、保存精美,为国内外罕见[39]。

1.2.6 侏罗纪燕辽生物群

侏罗纪是中生代生物界大发展时期。随着自早侏罗世大西洋的初步开裂,海洋范围不断扩大,全球气候进一步增温;环太平洋地区板块构造运动使内陆盆地大量增加,河湖密布和气候的温暖湿润为动、植物的繁盛创造了有利条件;在亚洲大陆上,早侏罗世晚期至中侏罗世早期是又一个重要的成煤时期。

侏罗纪初期的辽宁,总体上继承了晚三叠世的温暖潮湿气候,森林仍然繁茂,形成了以早侏罗世北票组为代表的含煤地层,是辽宁"第2次主要成煤期"[39]。但由于剧烈的火山活动等地质地理条件变化,加之生物自身的演化,自中侏罗世起,辽宁及其邻区生物群面貌发生了重大改变,形成了"燕辽生物群"。燕辽生物群是生活在距今约1.8亿—1.5亿年侏罗纪中、晚期的一个生物群,主要分布于我国辽西及其邻区冀北和内蒙古东南

一亿多年前的"肉食小强"——蟑螂

现生蟑螂——美洲大蠊(*Periplaneta americana*)

化石蟑螂——端色强壮蛇蠊(*Fortiblatta cuspicolor*)产于侏罗纪—白垩纪,已知最早距今1.65亿年。

蟑螂,拥有强大生命力和古老历史,存活了1亿6千多万年。地史中的一类蟑螂称"蛇蠊",被戏称"肉食小强",繁盛于侏罗纪和白垩纪,数量及种类繁多,它们体型狭长,复眼发达,行动灵活,具有良好视力和宽阔视野。口器位于头部前端,上颚强壮,前足发达并长满硬刺,便于攻击和捕获猎物,主要以昆虫为食,具"食肉"特性。这明显不同于现生蟑螂,今天所有的蟑螂均"从良"吃素了,它们专吃水果、面包碎屑、绵毛皮革制品或书纸,甚至肥皂、垃圾及动物尸体等,就是不吃活动物的肉。

那么,为什么古老的蟑螂经过一亿多年的演变从"肉食"改为"素食"了呢?

辽西燕辽生物群景观(据张宜,2015)

部沿燕山一带;其更大的分布范围可东延辽东,西北至河西走廊、中亚东北部以及俄罗斯外贝加尔一带。

"燕辽生物群"这个名字最初来自于我国古昆虫学家洪友崇(1983)命名的"燕辽昆虫群"[9],其原含义主要指冀北的九龙山组和辽西海房沟组所产的昆虫动物群,时代为中侏罗世。此后,我国年轻的古昆虫学家任东(1995)将其含义扩大为"燕辽动物群"[33],层位包括中侏罗世海房沟组(九龙山组)和髫髻山组,并将该动物群的层位下延至冀北的"门头沟组"和辽西的北票组,上限延伸到土城子组。2011年,孙革等提出将辽西及冀北等地区侏罗纪生物群统称为"侏罗纪燕辽生物群"[39],但经新一轮研究考虑,为突出这一生物群在中、晚侏罗世的特征、并考虑这一生物群最初命名的由来,"燕辽生物群"的涵义还是以仅限于辽西及冀北等地的中、晚侏罗世的生物群为宜。为此,本书提出的"侏罗纪燕辽生物群"系指生活于辽西及冀北等地的中、晚侏罗世生物群,不再包括这一地区早侏罗世北票

能吸恐龙血液的跳蚤——巨型跳蚤

现生跳蚤(Pulex irritans)是一种小型、无翅、善于跳跃的寄生性吸血昆虫,寄生于人类、哺乳动物及鸟类的体表,用锋利的口器吸食寄主的血液,同时将自身的病菌传给宿主。恐龙时代的巨型跳蚤(giant fleas)体型巨大,体长超过2厘米,大约是现生跳蚤的10倍;没有翅膀,触角很短,长长的口器、具栉状刚毛的足以及体表发育密集向后的鬃毛,揭示其适于寄生于具毛的脊椎动物体表,如哺乳动物和带毛恐龙;它们长而尖锐且具锉状小齿的口器可穿透恐龙较为坚韧的皮肤,吸血而食。巨型跳蚤化石由我国青年古生物学专家黄迪颖等2012年发现于内蒙古宁城道虎沟、距今约1.65亿年的中侏罗世地层,成果轰动世界。

那么,为什么跳蚤经过一亿多年的演化会变得越来越小了呢?

现生跳蚤

化石跳蚤——巨型跳蚤

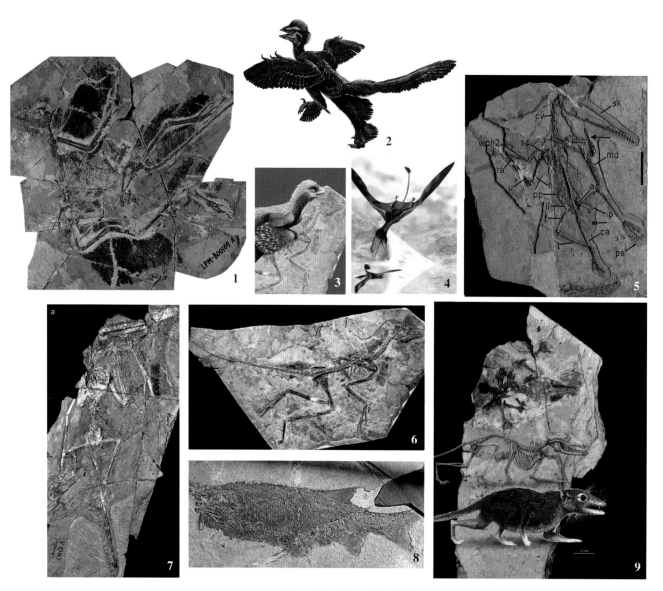

燕辽生物群部分代表性动物
1,2. 赫氏近鸟龙；3,6. 曙光鸟（*Aurornis*）；4,5. 达尔文翼龙（*Darwinopterus*）雌性个体；
7. 悟空翼龙（*Wukongopterus*）；8. 褶鳞鱼（*Ptycholepis*）；9. 中华侏罗兽。

组生物群的内容。

燕辽生物群大体上可分为中侏罗世部分（包括中—晚侏罗世之交时期，也称海房沟—蓝旗—髫髻山生物组合）和晚侏罗世部分（土城子生物组合）等两部分。整个生物群的类群已达20多个门类上千种。该生物群以中侏罗世（包括中—晚侏罗世之交）辽西建昌、冀北丰宁及内蒙古宁城道虎沟地区的化石最具代表性，以带羽毛恐龙、早期哺乳动物、翼龙、植物、昆虫及鱼类等最具特色。其中，特别是迄今世界最早的带羽毛恐龙赫氏近鸟龙和迄今最早的真兽类哺乳动物中华侏罗兽的发现，为全球鸟类起源和真兽类哺乳动物起源等研究作出突出贡献。至晚侏罗世（以土城子组为代表），生物群组成已大大衰减，但

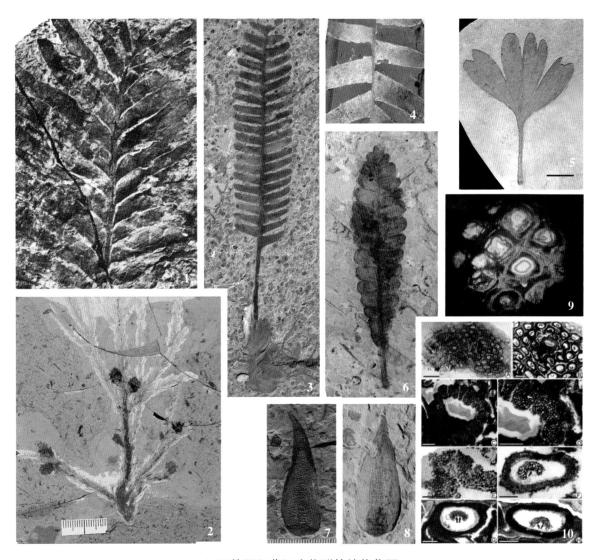

辽西侏罗纪燕辽生物群的植物化石

1. 庞特篦羽叶（*Ctenis pontica*）；2. 燕辽杉；3. 尼尔桑（*Nilssonia*）；4. 侧羽叶（*Pterophyllum*）；
5. 银杏；6. 异羽叶；7，8. 苏铁鳞片；9. 北票南洋杉雌球果（据 Zheng et al., 2008）；
10. 北票阿什茎（*Ashicaulis beipiaoensis*）解剖（据 Tian et al., 2014）。

也曾发现大型蜥脚类恐龙骨骼及其足迹以及鱼类化石等[45]。

燕辽生物群的总体特征为：生物分异度明显加大，恐龙、哺乳动物的一些新的分类群突兀出现（如爬行动物首次出现带羽毛恐龙、哺乳动物首次出现有胎盘的真兽类等），翼龙类及昆虫异常繁盛，古植被面貌也有较大变化。中侏罗世早中期的气候较潮湿和温暖，但具有季节性的干旱；到中侏罗世晚期，气候逐渐转向干热，造煤作用已完全停止；至晚侏罗世，气候已变得炎热而干旱，可能只是在局部沙漠绿洲有一些生物生存。

燕辽生物群的植物化石包括早期（海房沟期）、中期（蓝旗—髫髻山期）和晚期（土城子期）三部分。早期部分的植物大化石以简单锥叶蕨—裂叶爱博拉契蕨（*Coniopteris simplex-*

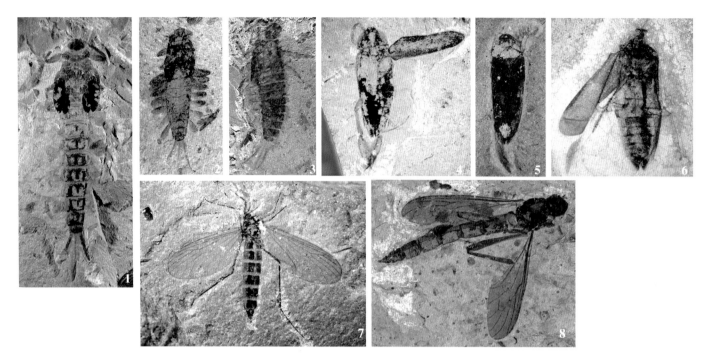

燕辽生物群中的昆虫化石

1. 群居蜉(*Fuyous gregarius*); 2. 湖山头蜉(*Shantous lacustris*); 3. 残遗暗蜉(*Furvoneta relicta*);
4. 火山道虎沟划蝽(*Daohugocorixa vulcanica*); 5. 真九龙山划蝽(*Jiulongshanocorixa genuina*);
6. 中华燕辽划蝽(*Yanliaocorixa chiensis*); 7. 赤峰长室冬大蚊(*Tanyochoreta chifengica*);
8. 罕见新始虻(*Novisargus rarus*)。1—4、7、8均产于道虎沟;
5产自九龙山;6产自海房沟。(据张俊峰,2015*)

Eboracia lobifolia)组合为代表,孢粉化石以似抄椤孢—阿赛勒特孢—似紫萁孢(*Cyathidites-Asseretospora-Osmundacidites*)组合为代表,主要产于海房沟组及其相当地层,已发现48属124种[60,30,31],该组合以出现本内苏铁类带生殖器官的异羽叶(*Anomozamites*)[98]、威特里奇(*Weltrichia*)及大量松柏类燕辽杉(*Yanliaoa*)为特色;海房沟组有薄煤层或夹煤线。在辽东,该组合以本溪田师傅中侏罗世大堡组植物化石(33属70余种)为代表,并形成一些工业煤层[67],反映此间辽东的气候比辽西更加温暖潮湿。中期部分植物大化石以篦羽叶—中国小威廉姆逊(*Ctenis-Williamsoniella sinensis*)组合为代表,孢粉化石克拉梭粉的含量比海房沟组增加,主要产于辽西北票长皋蛇不歹及建昌玲珑塔等地的髫髻山组,时代为中侏罗世晚期[60],该组合以盛产木化石为特色,如郑少林等发现的李氏木(*Lioxylon*)、田宁等发现的紫萁科茎干化石阿什茎(*Ashicaulis*)及似落羽杉型木(*Protaxodioxylon*)等[41,42,61,88,89];此外,郑少林等还发现北票南洋杉(*Araucaria beipiaoensis* Zheng, et al.)雌球果,表明现今仅生长在南半球热带地区的南洋杉在地质历史曾在北半球生存过。辽西髫髻山组是

---

* 张俊峰. 2015. 辽西燕辽生物群和热河生物群的昆虫化石(未刊)

一套中性的火山喷发岩系,含有4个沉积岩夹层,各夹层均属于河流相沉积,但在火山喷发间歇期植被都很繁茂,其中尤以本内苏铁类占有绝对优势,苏铁类以羽叶宽大的篦羽叶特别繁盛为特征,表明这两类植物特别适应当时的气候环境。辽东田师傅三个岭组所产植物化石也属于这个组合。当然,上述两部分的植物也都含大量落叶类型的银杏类(如 *Ginkgo*)及茨康类(如 *Czekanowskia*),反映这一时期气候存在季节性的变化。

燕辽生物群的昆虫化石已发现有17个目、近百个科、数百个种,以膜翅目寄生蜂类和双翅目蚊类最丰富(二者共超过100余种)[9,10,21,32,33]。传统上认为,海房沟组、九龙山组和道虎沟层中的水生昆虫优势种组合基本相同[9,33],即以西伯利亚中蜉(*Mesobaetis sibirica* Brauer et al., 1889)、古珠蜉(*Mesoneta antiqua* Brauer et al., 1889)和中华燕辽划蝽(*Yanliaocorixa chiensis* Lin, 1976)等为代表;但据张俊峰等最新研究[95~97]表明,海房沟组顶部的水生昆虫优势种组合是残遗暗蜉—中华燕辽划蝽(*Furvoneta relicta-Yanliaocorixa chinensis*);辽西邻区宁城道虎沟层同时期水生昆虫优势种组合为群居蜉—湖山头蜉—火山道虎沟划蝽(*Fuyous gregarius-Shantous lacustris-Daohugocorixa vulcanica*),其组合特征并不完全相同;但根据生物地层学对比,海房沟组和道虎沟层与哈萨克斯坦晚侏罗世的卡拉套昆虫群非常相似,具有高度的可比性,与俄罗斯伊尔库茨克的乌斯契—巴列伊盆地契林霍夫组的昆虫组合也可以进行对比。

晚侏罗世土城子期化石主要包括恐龙、鱼类、叶肢介、介形类、昆虫、双壳类及植物

一亿多年前的"歌唱家"——蝉

人们万万没想到"歌唱家"早已在一亿多年前的辽西北票出现,这就是昆虫界著名的"辽蝉"。

蝉俗称"知了",以树汁为食,幼虫栖息在土中,吸食树根汁液,是对树木有害的昆虫。

会唱歌的蝉是雄蝉,"乐器"位于肚子下方,像蒙上一层鼓膜的大鼓,鼓膜受振动而发出声音。雄蝉一到夏天,为引诱雌蝉交配,便站在树上"知了"个没完,"歌声"传出很远,许多雄蝉一起叫,仿佛是一个合唱团。蝉鸣还可预报天气,所谓"知了鸣,天放晴","蝉儿叫叫停停,阴雨将要来临";在炎炎夏日里,蝉鸣也预示炎热天气将持续。

一亿多年前的雄蝉已经是"歌唱家",那么雌蝉当时是怎样与他一起合唱的呢?

北票辽蝉(*Liaocossus beipiaoensis*)

辽宁北票哑巴沟土城子组大型蜥脚类恐龙化石
1. 化石发掘现场；2. 恐龙学家董枝明（左2）在现场指导发掘工作。

等。该期植物化石属于燕辽生物群植物组合的晚期部分，大化石主要以短叶杉—尖叶杉（*Brachyphyllum-Pagiophyllum*）组合为代表，已发现39种（包括5种木化石）[65,61]；孢粉化石以克拉梭粉—四子粉（*Classopollis-Quadraeculina*）组合为代表，掌鳞杉科的克拉梭粉的含量高达57%—82.6%[30,31]。土城子组主要是一套灰绿或紫红色的交错层砂岩沉积，属风成沙丘或旱谷堆积，在沙漠绿洲的水源附近有一些植物生长。晚侏罗世可能存在气候升温事件，呈现干旱气候，只是能抵御干热的植物（如掌鳞杉科等）数量较多。

土城子期脊椎动物以杨氏朝阳龙—热河恐龙足印（*Chaoyangsaurus yaungi-Jehosauripus*）组合为代表，主要见于辽西朝阳地区。2015年夏，在北票大板镇孤家子乡哑巴沟村又

恐龙时代的"蝴蝶"——丽蛉

 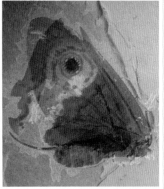

现生蝴蝶——黑猫头鹰环蝶
(*Caligo atreus*)
（图片据SmithsonianShare）

化石蝴蝶——丽蛉

丽蛉（*Oregramma illecebrosa*），繁盛于中侏罗世至早白垩世一类已灭绝的脉翅目昆虫，体型较大，前后翅宽广，且密密麻麻地布满了细小的横脉，身体和翅常具有较密的刚毛，翅面具有明显的形态多样的翅斑、条纹或眼斑，表面上看似大型的蝴蝶，因此常被人误认为是"侏罗纪的蝴蝶"。事实上它要比真正的蝴蝶早出现至少6000万年。丽蛉翅面上的这些翅斑、条纹或眼斑，用以隐藏自己或者模拟脊椎动物的眼睛，躲避敌害。特别是眼斑，休息时是隐藏的，一旦受到惊扰，展翅突然暴露眼斑，对捕食者产生威慑作用，吓退捕食者。

发现较大型蜥脚类恐龙,体长大于10米。在新宾小东沟组发现有大量的褶鳞鱼类(Ptycholepidae)化石。土城子期叶肢介化石以假线叶肢介—北票叶肢介—中雕饰叶肢介(*Pseudograpta-Beipiaoestheria-Monilestheria*)组合为代表[45]。介形类化石早期以小怪介—曼特尔介—达尔文介(*Cetacella-Mantelliana-Darwinula*)组合为代表,已发现5属35种;晚期以准噶尔介—曼特尔介—斯特内斯措姆介(*Djungarica-Mantelliana-Stenestroemia*)组合为代表,已发现8属25种[52]。

1.2.7 早白垩世热河生物群

热河生物群(Jehol Biota)是约1.4亿—1.2亿年前生活在亚洲东部地区(包括我国东北部、蒙古、俄罗斯外贝加尔、朝鲜等)的一个古老的生物群;以我国辽西义县—北票—凌源等地区为最主要产地。该生物群最初曾以东方叶肢介—三尾拟蜉蝣—狼鳍鱼(*Eosestheria-Ephemeropsis trisetalis-Lycoptera*)为典型代表[8]。近20年来,辽西热河生物群大量珍稀化石,如带毛恐龙中华龙鸟[16]、具4个翅膀的恐龙小盗龙[91]、原始鸟类孔子鸟[12]、会滑翔的蜥蜴赵氏翔龙(*Xianglong zhaoi*)[39]、早期真兽类哺乳动物始祖兽(*Eomaia*)[79],以及迄今"最早的花"辽宁古果[85]和中华古果[86]等。这一特有的中生代生物群已成为国际古生物学界和相关科学界关注的热点,有关热河生物群一些重要化石发现为研究鸟类起源与早期演化、真兽类起源、被子植物起源及昆虫与有花植物的协同演化等重大理论问题提供了宝贵的化石依据。

纵观近20年来有关热河生物群的研究进展,早白垩世热河生物群的生物多样性尤其

一亿多年前的"空中小猎手"——蜻蜓

假如不是在辽宁古生物博物馆亲眼见到产于辽西的化石,谁敢相信距今一亿多年前,蜻蜓——这些"空中小猎手"早已经活跃在温暖、潮湿的森林和湖边的草丛中,成为辽西"热河昆虫群"中一群婀娜多姿的成员。

人们今天看到的蜻蜓,是昆虫界的飞行专家,飞行技术高超,快飞、慢飞,甚至倒飞,无所不能,飞行速度每小时可达30千米。蜻蜓还具有非常好的视力,复眼由1—3万个小眼睛组成,占据头部的大部分;头部还可以180°转动,使它具有更宽阔的视野,飞行中能随时发现猎物。一旦发现猎物,蜻蜓会像离弦的箭一样迅速冲向目标,用长满尖刺的足去捕获猎物,之后,饱餐一顿,更有甚者边飞边吃。

蜻蜓的确是昆虫界名副其实的空中猎手。那么,一亿多年前的"中国蜓"也有这样的本领吗?

化石蜻蜓——多室中国蜓(*Sinaeschnidia cancellosa*)产于义县组,距今约1.25亿年。

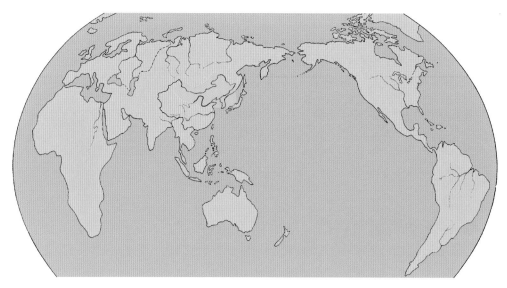

热河生物群的古地理分布示意图(绿色示)(据孙革等,2011)

令人瞩目。迄今在热河生物群中已发现恐龙、鸟类、翼龙类[90]、两栖类、龟鳖类、蜥蜴类、哺乳类、鱼类、昆虫、蜘蛛、双壳类、腹足类、介形类、叶肢介、虾类、鲎虫类、植物(包括木材化石和孢粉)、轮藻等近20多个门类千余种化石[75,77,78,81,85,86,90,99,101,102]。仅脊椎动物已报道至少有129属155种(包括鸟类约38属44种,恐龙约31属37种,翼龙约21属21种,哺乳动物约14属16种)。介形类、双壳类、叶肢介、腹足类等无脊椎动物约100属290种;昆虫类约327属467种;植物(包括孢粉)约90属150余种。上述化石主要见于义县组,部分见于九佛堂组。

热河生物群生态复原图(据孙革等,2011)

1.2.7.1 恐龙

辽西热河生物群中的恐龙,特别是带羽毛的小型兽脚类恐龙,为鸟类起源和羽毛早期演化等研究提供了宝贵的化石证据。在辽西最早发现的保存有原始羽毛的兽脚类恐龙中华龙鸟,以及带羽毛兽脚类原始祖龙(Protarchaeopteryx)和尾羽龙(Caudipteryx)等的发现均为鸟类由兽脚类恐龙演化而来提供了重要证据。镰刀龙类的北票龙(Beipiaosaurus)和奔龙类的中国鸟龙(Sinornithosaurus)的纤维状原始羽毛表明,羽毛不再是鸟类专有的特征。个体较小的奔龙类——小盗龙,其第一趾位置低、且所有趾爪的钩曲度较大等特征为飞行的树栖起源假说提供了证据;其4个翅膀和前后肢均发育着不对称的羽毛等复杂结构,表明这些带毛恐龙已具有一定的飞行功能,且显示早期鸟类演化中可能经历了"四翼"阶段。此外,近年

辽西热河生物群中的主要恐龙
1,2. 中华龙鸟;3,4. 尾羽龙;5. 小盗龙;
6. 东北巨龙(左)和辽宁巨龙(右)。

来的新发现表明,辽宁西部在这一时期也同时存在着形体巨大的植食性恐龙,如辽宁巨龙（*Liaoningotitan*, MS）和东北巨龙（*Dongbeititan*）等。

1.2.7.2 鸟类

热河生物群中的鸟类已具有很高的分异度,迄今已发现的化石超过38属44种[107],代表了鸟类历史上第一次大规模辐射演化。此次辐射演化不仅表现在分类上,而且在形态、个体大小、飞行能力、习性等方面也有较大分异。其中,原始的基干鸟类孔子鸟及钟健鸟（*Zhongjianornis*）,迄今最大的鸟类会鸟（*Sapeornis*,反鸟类）及渤海鸟（*Bohaiornis*,反鸟类）,今鸟类的燕鸟（*Yanornis*）、建昌鸟（*Jianchangornis*）、叉尾鸟（*Schizooura*）及丽鸟（*Bellulia*）等均具代表性。热河生物群中的鸟类基本勾勒出了鸟类早期演化历程,即自原始的古鸟类（Archaeornithes）至反鸟类（Enantiornithes）再至今鸟类（Neoornithes）。

1.2.7.3 翼龙类

辽西热河生物群中已发现了至少21个翼龙属种,代表着早白垩世时期一个重要的具有高分异度的翼龙类群,对研究翼龙目中两大类群（喙嘴龙类和翼手龙类）的更替、翼手龙类的辐射演化以及与鸟类在生态位上的竞争等,都有着重要意义。热河生物群中的翼龙化石多以无齿类翼龙为主,这一时期的翼龙在形态特征及生活习性等方面都展示出很大

季强,江苏南通人,古生物学家,中国地质科学院研究员。1977年毕业于南京大学,1990年于德国森肯堡研究所获博士学位,曾任中国地质博物馆馆长。1996年他与姬书安研究员合作发现热河生物群最早的带毛恐龙中华龙鸟,有力地推动了全球鸟类起源研究和热河生物群研究。他还率队发现尾羽鸟及中华侏罗兽等,为辽宁及全球中生代生物群研究作出突出贡献。

季强（1951— ）

徐星,新疆伊犁人,古脊椎动物学家,中国科学院古脊椎动物与古人类研究所研究员。1992年毕业于北京大学,2002年于中国科学院古脊椎动物与古人类研究所获博士学位。专长于恐龙研究。他首次发现小盗龙,首次命名近鸟龙等带羽毛恐龙,并提出恐龙演化的"四翼阶段"假说,有力地推动了全球鸟类起源研究,为辽宁及全球恐龙演化研究作出突出贡献。

徐星（1969— ）

热河生物群鸟类

1,2,4. 孔子鸟；3,5. 会鸟(据周忠和等,2002)；6,7. 燕鸟(据周忠和等,2001)；8. 钟健鸟；9. 沈师鸟；10. 盛京鸟(*Shengjingornis*)；11. 翔鸟(*Xiangornia*)。

侯连海(1935—)

侯连海,山东单县人,古鸟类学家,中国科学院古脊椎动物与古人类研究所研究员。1961年毕业于兰州大学。1984年他与周忠和等首次发现我国中生代第一件鸟化石甘肃鸟,1995年首次发现原始鸟类孔子鸟,为我国辽西热河生物群古鸟类研究作出突出贡献。他撰写的《中国辽西中生代鸟类》等著作是我国及全球古鸟类研究的经典之作。

周忠和(1965—)

周忠和,江苏江都人,古鸟类学家,中国及美国科学院院士,中国科学院古脊椎动物与古人类研究所所长,国际古生物学会主席。1986年毕业于南京大学,1999年于美国堪萨斯大学获博士学位。1995年他与侯连海等首次发现原始鸟类孔子鸟,并系统研究了鸟类恐龙起源、早期鸟类的系统演化、生态分异及食性分化等,为我国古鸟类及热河生物群的综合研究等作出了重要贡献。

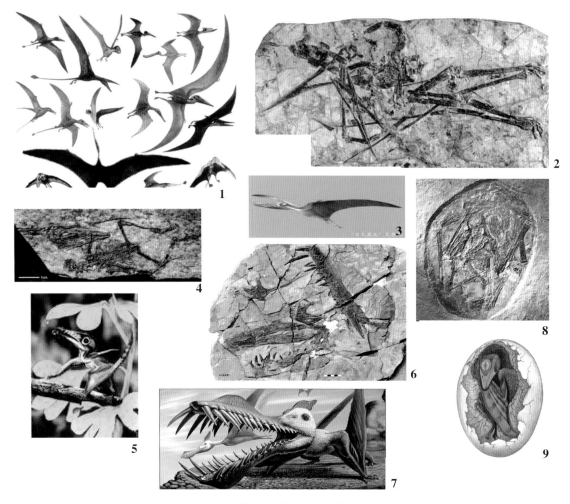

辽西热河生物群的翼龙类

1. 辽西翼龙多样性（据周长付，2015*）；2，3. 吉大翼龙；4，5. 最小的树栖翼龙——森林隐翼龙；6，7. "最大的翼龙"——辽宁翼龙（据汪筱林等，2003）；8，9. 翼龙的胚胎发育（据汪筱林等，2004）。

的分异，如两翼展宽可达5米的辽宁翼龙（*Liaoningopterus*）与翼展只有约25厘米的隐翼龙（*Nemicolopterus*，被视为"世界上最小的翼龙"）。此外，精美保存的翼龙胚胎化石首次证实了翼龙是以革质的软壳蛋进行繁殖的卵生动物。

1.2.7.4 其他动物

热河生物群中的其他动物化石也十分丰富。早期哺乳动物以迄今最早的有袋类动物沙氏中国袋兽（*Sinodelphys szalayi*）及早期有胎盘类哺乳动物攀援始祖兽（*Eomaia scansoria*）等为代表。两栖类包括无尾的蟾蜍类和有尾的蝾螈类，如天义初螈（*Chunerpeton*

---

* 周长付. 2015. 国土资源部专项《辽西早白垩世神龙翼龙超科的分类演化研究》中评估报告（未刊）

tianyiensis）。蜥蜴类中，迄今最早的会滑翔的蜥蜴赵氏翔龙最引人注目，该化石与现生的飞蜥相似，似也指示了辽西热河生物群的生态环境可能存在至少是具季节性的亚热带或接近亚热带雨林的环境。离龙类目前已发现至少有3种不同的生态适应类型：高度水生适应的长颈型的潜龙（*Hyphalosaurus*），半水生短吻型的满洲鳄（*Monjurosuchus*）和嬉水龙（*Philydosaurus*），以及长吻型的伊克昭龙（*Ikechosaurus*）等。热河生物群的鱼类化石也种类繁多，不仅包括小的狼鳍鱼（*Lycoptera*）和大的吉南鱼，长吻的原白鲟（*Protopsephurus*）和短吻的燕鲟，还包括肉食性的中华弓鳍鱼等，这些鱼类的繁盛可能为四足动物的繁衍提供了必要的食物资源。这些重要发现表明，热河生物群中的多门类脊椎动物在早白垩世时期均经历了重要的演化阶段。

热河生物群的无脊椎动物也十分丰富，包括腹足类、双壳类、叶肢介、介形类、蜘蛛、虾类和昆虫等近10个门类，它们为研究热河生物群的组成、协同演化以及其古地理环境等发挥了重要作用。热河生物群中的昆虫种类最为丰富，目前已发现至少143科327

热河生物群的翼龙

翼龙（pterosaur）是爬行动物中最早成功飞上天空的一个类群。与鸟类和蝙蝠不同，翼龙的膜质"翅膀"是依靠第4指骨（也称翼指骨）加长的指节为支撑和驱动来实现飞行。从三叠纪晚期的出现，到白垩纪末期的灭绝，翼龙类占据着中生代的天空约1亿5千万年，是绝对的空中霸主。

传统上，翼龙目被分为喙嘴龙亚目（Rhamphorhynchoidea）和翼手龙亚目（Pterodactyloidea）。前者较为原始，以颈短、尾长、掌骨短、第Ⅴ趾长为特征，主要分布于三叠纪晚期至侏罗纪末期；后者较为进步，以颈长、尾短、掌骨长、第Ⅴ趾强烈缩短或退化为特征，主要分布在晚侏罗世至晚白垩世。

热河生物群的翼龙化石非常丰富，以翼手龙类最为繁盛，已发现约30个属种，化石不仅骨骼保存精美，而且翼膜、毛发等软组织印痕都有保存；特别是含胚胎的翼龙蛋化石确切证实了翼龙为卵生动物，解决了困惑古生物学家100多年的有关翼龙类如何繁殖的难题。热河生物群翼龙生态上高度分异，如滤食性的格格翼龙、飞龙和莫干翼龙，食鱼性的辽宁翼龙、鬼龙和伊卡兰龙，肉食性的努尔哈赤翼龙、帆翼龙和红山翼龙，植食性的中国翼龙、华夏翼龙和森林翼龙等。

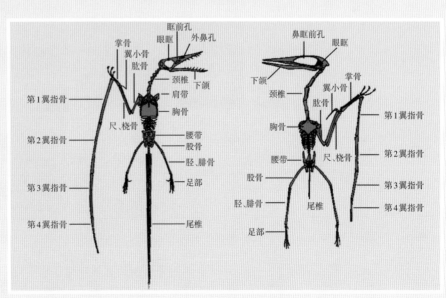

喙嘴龙类（左）和翼手龙类（右）的骨骼对比图（据吕君昌等，2006）

属467种[107]，以含三尾拟蜉蝣（*Ephemeropsis trisetalis*）和长肢裂尾甲（*Coptoclava longipoda*）为特色。热河生物群的昆虫群主要包括义县组昆虫组合和九佛堂组昆虫组合，前者主要以蚊蝇、蜻蜓、甲虫、蜉蝣、蜂和蜡组合为代表，后者主要以蚊蝇类、甲虫、蜉蝣和蜂类居多。如在义县昆虫组合发现的、被誉为"三栖明星"的中国蜓（*Sinaeschnidia*）、"爱唱歌的蝈蝈"鸣螽（*Habrohagla*）、"实力唱将"辽蝉（*Liaocossus*）、"天敌昆虫"巴依萨蛇蛉（*Baissoptera*）以及"植物的红娘"细蜂（*Gurvanotrupes*）等，均独具特色[39]。特别需要指出的是，在热河生物群的昆虫化石中，已发现喜花昆虫[如花蚤类（Mordellidae）、花蝽类（Anthocoridae）、泥蜂类（Sphecidae）和侏罗原网翅虻（*Protonemestrius jurassicus*）等]的出现[81]，这些喜花昆虫为研究古昆虫与早期被子植物的协同演化提供了重要证据。目前在义县组的喜花昆虫

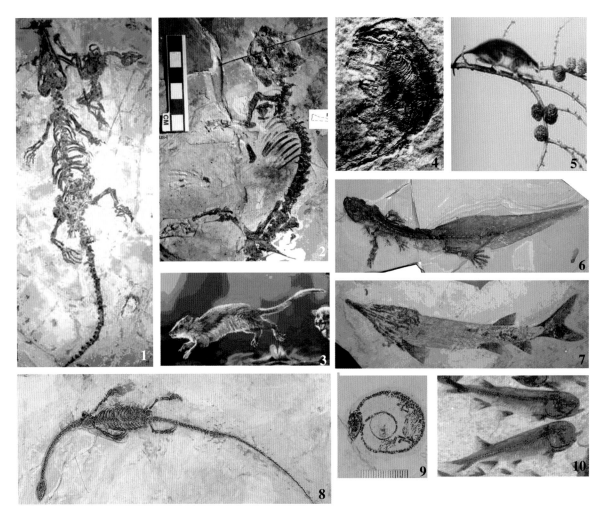

热河生物群的其他脊椎动物
1. 伊克昭龙；2,3. 皱纹齿兽（*Rugosodon*）；4,5. 攀援始祖兽；6. 天义初螈；
7. 原白鲟；8,9. 潜龙；10. 狼鳍鱼。

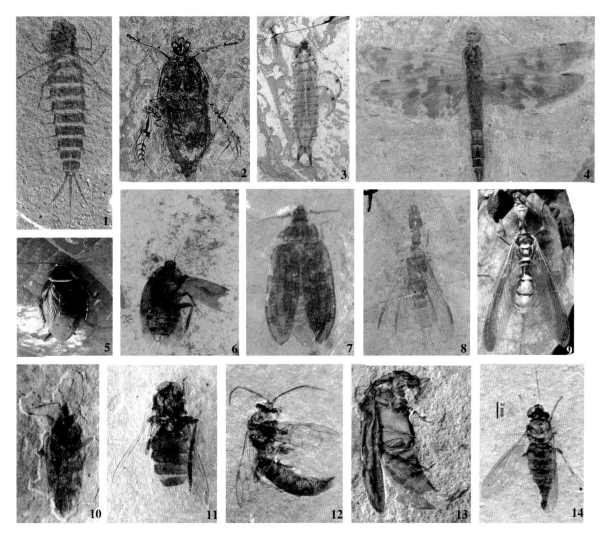

热河生物群的昆虫

1. 三尾拟蜉蝣；2,3. 长肢裂尾甲，3 为幼虫；4. 多室中国蜓；5,6. 蜚蠊 (Blattidae)，5 为现生；
7. 波纹眼甲 (Amblomma)；8. 巴依萨蛇蛉；9. 现生蛇蛉 (Climaciella)；10. 网翅虻类；
11. 花蚤类；12. 花蜂类；13. 泥蜂类；14. 侏罗原网翅虻。(据任东，1998)

中已发现至少有两种类型口器用以适应不同的长花管状和开放型的花朵（张俊峰，2015*）。

1.2.7.5 植物

热河生物群植物化石十分丰富，迄今已发现大化石 50 余属近百种。该植物群最突出的特征是：(1) 松柏类占优势（32 种，占 36.4%），其常绿属种（如南洋杉、柏型枝、

---

* 张俊峰. 2015. 辽西燕辽生物群和热河生物群的昆虫化石（未刊）

辽宁古果的模式产地
北票黄半吉沟

短叶杉、坚叶杉及似罗汉松等）占一半以上，反映温暖并可能地处低矮山地；（2）本内苏铁类占一定比例（8种），似反映温湿度均偏高；（3）买麻藤类数量多，可能具有季节性旱生特征；（4）茨康类及银杏类有一定比例（共占约13.6%），反映有季节性变化；（5）首次发现早期被子植物出现；（6）地方性属种约占一半以上。上述特征综合反映植物群很可能地处山地间的河湖附近、具有适温热兼季节性干旱的植被性质，并具有较强的地方性特色[39]。

热河生物群的早期被子植物以古果属（Archaefructus，包括辽宁古果、中华古果等）为主要代表，与之同期出现的早期被子植物还有十字里海果（Hyrcantha decussata）[71]等。"古果"的生殖枝上螺旋状着生数十枚蓇葖果，由心皮对折闭合而成，果实（心皮）内包藏着数粒种子（胚珠），具单沟状花粉。其茎枝细弱，叶子细而深裂，根不发育，未见花瓣和花萼，属于典型的水生草本被子植物。根据"古果"等在早期被子植物中的原始性，以及地质和地理分布等特征，孙革等（1998，2002）首次提出"被子植物起源的东亚中心"假说，并提出被子植物存在水生起源的可能性。上述成果有力地推动了我国及全球被子植物起源与早期演化研究，并被国内外教科书和许多著名专家学者引用；美国著名被子植物专家Doyle（2012）已将"古果"属的出现、连同其所产出地层义县组作为全球早期被子植物演化的底界[72]。

1.2.8 早白垩世阜新生物群

阜新生物群是继早白垩世早—中期的热河生物群之后发展起来的一个早白垩世偏晚期的生物群，距今约1.1亿年，大体相当于阿普梯期—阿尔必早期。由于这一时期气候温暖潮湿，森林密布，植物繁茂，形成大规模的工业煤层，是辽宁"第3次主要成煤期"。阜新

热河生物群中的植物

早期被子植物：1,2. 辽宁古果；3—5. 中华古果及复原；6,7. 十字里海果及复原；8. 李氏果；9. 古果生态复原。伴生植物：10. 长叶似木贼；11. 夏家街蕨；12. 热河拟节柏；13. 尖山沟威廉姆逊；14. 薄氏辽宁枝；15. 似管状叶；16. 古尔万果；17,18. 北票果；19. 卵形毛籽；20. 羽状枞型枝；21. 热河裂鳞果；22. 陈氏似麻黄。

生物群主要产于辽宁阜新的沙海组和阜新组,与之相当的同期生物群主要位于铁岭西部的铁法地区等[39]。

"阜新生物群"这个名称是由王五力(1987)首先提出、并从广义的热河生物群中划分出来的[43]。阜新生物群由丰富的哺乳类、鱼类、叶肢介、双壳类、腹足类、介形类、昆虫及植物等组成。该生物群主要产于辽西的沙海组和阜新组及其相当层位,广泛分布于中国北方及俄罗斯远东和日本等邻区。

阜新生物群的植物化石最为丰富。早在20世纪30年代初,我国古植物学家斯行健根据阜新露天矿阜新组的标本建立了高腾刺蕨(*Acanthopteris gothani* Sze)新属种,成为阜新植物群及东北、华北早白垩世植物群重要分子和对比标准[34]。据郑少林(2011*)研究,阜新植物群至少由93种植物大化石组成,以高腾刺蕨—中国蕉羽叶—七弦琴篦羽叶(*Acanthopteris gothani-Nilssonia sinensis-Ctenis lylata*)组合为代表。其早期亚组合以阜新组的高德层至孙家湾层所产化石为代表,晚期亚组合以阜新组上部的水泉层所产的化石为代表。陈芬等[2]对阜新组植物化石的系统研究,孟祥营等[26]描述的亚洲杉木(*Cunninghamia asiatica*),郑少林、张武[66]研究的阜新侧羽叶(*Pterophyllum fuxinense*),商平及王自强报道

斯行健(1901—1964)

斯行健,浙江诸暨人,国际著名古植物学家,中国科学院学部委员(院士)。1926年毕业于北京大学,1933年于德国柏林洪堡大学获博士学位,曾任中国科学院南京地质古生物研究所第二任所长,是中国古植物学奠基人。他编著的《中国古生代植物图鉴》、《中国中生代植物》和《陕北中生代延长层植物群》是我国古植物学的经典著作。他率先指出中国中生代植物群演替规律并提出中国中生代陆相地层划分方案,为中国古植物学的发展作出卓越的贡献。

陈芬(1930—2011)

陈芬,天津人,古植物学家。1952年毕业于北京大学,前北京地质学院(现中国地质大学)教授。她首次详细研究了辽宁阜新盆地及铁法盆地早白垩世植物群;发表了《北京西山侏罗纪植物群》等重要专著。她编著的《古植物学》等教科书曾培养了中国几代古生物学及地质工作者。她为中国中生代植物及地层学研究作出了重要贡献。

---

* 郑少林. 2001. 辽宁中生代植物群(未刊)

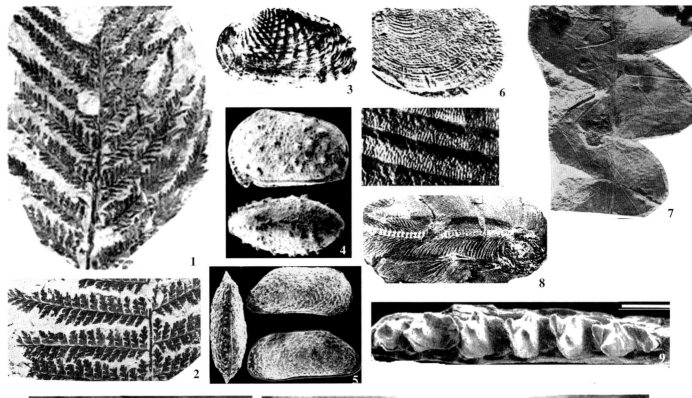

阜新生物群化石及阜新露天矿

1. 高腾刺蕨（*Acanthopteris gothani*）；
2. 瓦氏锥叶蕨（*Coniopteris vachrameevii*）；
3. 中华日本蚌（*Nippononaia sinensis*）；
4. 义县乌鲁威里女星介［*Cypridea*（*Ulwellia*）*insienensis*］；
5. 阜新始似玻璃介（*Eoparacandona fuxinensis*）；
6. 固阳内蒙古叶枝介（*Neimongolestheria guyangensis*）；
7. 七弦琴篦羽叶（*Ctenis lyrata*）；
8. 常氏海州鱼（*Haizhoulepis changi*）；
9. 鹿间明镇古兽（*Mozomus shikamai*）；
10. 恐龙足迹；
11. 阜新海州露天矿。

阜新生物群命名人王五力

的辽西鱼网叶(*Sagenopteris liaoxiensis*),以及厉宝贤对银杏类化石等的研究,基本上反映了这一植物群的总体面貌。阜新生物群的植物化石反映了温暖潮湿的北温带气候环境,适合于松柏类和银杏类等高大乔木型森林的发展,为当时的成煤提供了充足的物质来源。阜新植物群的孢粉化石主要包括三个组合(自早至晚):(1)沙海组组合,辽西孢—刺毛孢—克拉梭粉(*Liaoxisporis-Pilosisporites-Classopollis*)组合;(2)阜新组下部组合,刺毛孢—有突肋纹孢—三孔孢(*Pilosisporites-Appedicisporites-Triporolets*)组合;(3)阜新组上部(水泉层)组合,三角孢—无突肋纹孢—有突肋纹孢(*Deltoidospora-Cicatricosisporites-Appendicisporites*)组合[31]。

阜新生物群的脊椎动物主要包括哺乳类明镇古兽(*Mozomus*)、远腾兽(*Endotherium*)等。恐龙化石包括亚洲龙(*Asiatosaurus*)、黑山恐龙蛋(*Heishanoolithus*)和恐龙足迹化石等。鱼类化石有固阳鱼(*Kuyangichthys*)、薄鳞鱼类(Leptolepidae)及常氏海州鱼(*Haizhoulepis changi*)等[24,43]。无脊椎动物主要包括叶肢介、双壳类、介形类及腹足类等,十分丰富。叶肢介以延吉叶肢介—假瘤膜叶肢介(*Yanjiestherites-Pseudestherites*)组合为代表,但仍有 *Neimengolestheria* 等分子的存在,与吉林延边早白垩世晚期(阿尔必期)大砬子组中的延吉叶肢介(*Yanjiestheria*)群尚有一定区别[43]。双壳类化石以类三角蚌—褶珠蚌—日本蚌(*Trigonioides-Plicatounio-Nippononaia*)为代表,简称"TPN动物群"。介形类早期组合以义县乌鲁威里女星介—青河门湖女星介—近球状原微星介[*Cypridea* (*Ulwellia*) *ihsienensis-Limnocypridea qinghemenensis-Protocypretta subglobosa*]组合为代表,中期组合以微胀女星介—网状鳍女星介—小疹曼特尔介[*Cypridea* (*Cypridea*) *tumidiuscula-Pinnocypridea dictyodroma-Mantelliana papulosa*]组合为代表,晚期组合以球状假伟星女星介—栋梁玻璃介—阜新始似玻璃介[*Cypridea* (*Pseudocypridina*) *globra-Candona? Dongliangensis-Eoparacandona*]组合为代表[52,53]。昆虫化石以昼蜓—白垩沫蝉(*Hemeroscopus-Cretocercopis*)为代表[43]。

在辽西及辽北,阜新组之上不整合地覆盖一套以红色为主的沉积,称孙家湾组,时代可能为早白垩世最晚期(晚阿尔必期)至晚白垩世早期(赛诺曼期),其中有恐龙等化石发现。考虑到化石报道较少,其生物群在本书暂不单列。

1.2.9 古近纪抚顺生物群

古近纪初始,恐龙已灭绝,生物界开始了新的复苏。辽宁的古近纪生物化石十分丰富,主要反映在抚顺生物群(以大化石为主)和辽河油田的微体生物群。本书将这一生物群统称为"古近纪抚顺生物群",时代为古新世—始新世,距今约60—36Ma。抚顺生物群迄今已发现大量植物、鱼、昆虫、叶肢介、介形类、龟及小型的哺乳动物等十余个门类的化

抚顺生物群的植物化石

1. 抚顺苏铁（*Cycas fushunensis*）；
2. 中国沙巴榈（*Sabalites chinensis*）；
3、4. 二列水杉（*Metasequoia disticha*）；
5. 弧脉荚（*Viburnum speciosum*）；
6. 桤（*Alnus schmalhausenii*）
（2、5、6据张志诚，1980）；
7. 混杂副桤木粉（*Paraalnipollenites confuses*）
（据孙湘君等，1980）。

抚顺野外工作（2006）

石;昆虫化石常保存于精美的琥珀之中;植物化石以被子植物为主,也出现大量松柏类(如水杉、红杉)等高大植物;被子植物沙巴桐及苏铁等化石的发现,反映当时的气候曾相当炎热,植物茂盛,为形成大规模的森林、成为辽宁"第4次主要成煤期"[39]提供了充足的物质条件。与此同时,在辽宁中南部,辽河油田古近纪的微体生物群以介形类及藻类等的异常繁盛为特色,为石油的生成创造了有利条件(张立君,2011*)。

抚顺生物群最早由洪友崇所建[11],代表东亚大陆古新世—始新世(以始新世为主)一个独特的生物群。在抚顺盆地,化石主要产于古新统老虎台组及栗子沟组,始新统古城子组、计军屯组、西露天组及耿家街组。抚顺生物群的植物化石十分丰富,仅在始新世计军屯组已发现51属73种,以羽裂香蕨木—北极连香树—二列水杉(*Comptonia anderssonii-Cercidiphyllum arcticum-Metasequoia disticha*)组合为代表,以往亦曾称为"欧洲水松—二列水杉—抚顺漆"(*Glyptostrobus europaeus-Metasequoia disticha-Rhus fushunensis*)组合。该植物群被子植物占优势,约占84%;裸子植物居次,约占11%;植物群中既出现了古近纪

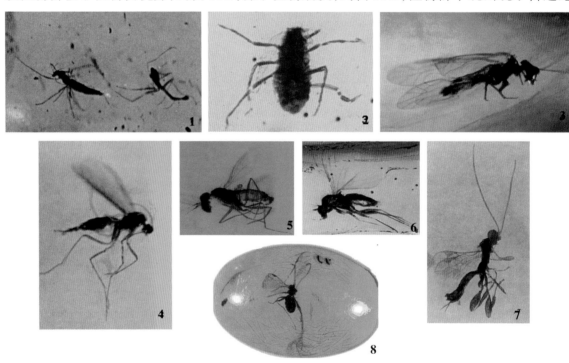

古近纪抚顺生物群的昆虫化石

1. 始新抚顺摇蚊(*Fushunitendipes eocenicus*)(左雌、右雄);2. 长形华夏蕈蚊(*Huaxiasciarites longus*);
3. 始新树蚜(*Silvaphis eocenica*);4. 古城子窄翅啮虫(*Stenopyerites guchengziensis*);
5. 抚顺始长足虻(*Eodolichopodites fushunensis*);6. 双尾中国蚤蝇(*Sinophorites bikerkosa*);
7. 窄形始冠蜂(*Eostephanites tenuis*);8. 始新短须蚁(*Curtipalpulus eocenicus*)。(据洪友崇)

---
* 张立君. 2011. 辽宁古近纪抚顺生物群和渤海湾生物群(未刊)

特有分子抚顺漆(*Rhus fushunensis*)等,又有大量现生科属,可与晚白垩世及新近纪植物群相区别。其中特别值得提及的是,在抚顺植物群中出现了中国沙巴桐[56]及抚顺苏铁[82]等热带—亚热带常绿植物,反映当时的气候曾相当炎热,这与始新世全球性升温事件相吻合。当然,该植物群中尚有一定数量的亚热带—暖温带落叶植物混生,表明该植物群可能属于亚热带常绿阔叶与落叶阔叶混交林;其水生植物槐叶萍(*Salvinia*)及黑三棱(*Sparganium*)等伴生,反映该植物群可能处于水体充沛的湖、河沿岸及其坡上。抚顺植物群的孢粉主要包括:(1)古新统组合,以高含量的三孔类花粉为特征,被子植物花粉占优势(达80%—96%),三孔类花粉以桦粉(*Betulaepollenites*,占17.3%)、副桤木粉(*Paraalnipollenites*,占17%—20%)及拟榛粉(*Momipites*,占9%—22%)为主;(2)始新统组合,以高含量的栎粉(*Quercoidites*)为特征,被子植物花粉占优势(约占60%—80%),其中,栎粉占40%以上[11,62]。

抚顺生物群的昆虫化石也十分丰富,主要产于古城子组煤层的琥珀和页岩中以及计军屯组下部的油页岩与细砂岩中,迄今已发现56科201属223种,分别隶属于蜉蝣目、䗛螳目、同翅目(蚜虫)、异翅目(或半翅目)、啮虫目、鞘翅目、双翅目和膜翅目等。该昆虫群以始新抚顺摇蚊—长形华夏蕈蚊—腹柄廖氏广肩小峰—抚顺卵头蚁—抚顺蓝绿象甲(*Fushunitendipes eocenicus-Huaxiasciarites longus-Liaoeurytomites petiolatus-Ovalicapito fushunensis-Hypomeces fushunensis*)组合为代表,双翅目的摇蚊科和华夏蕈蚊科的成员居首位[10,11]。抚顺昆虫群仅分布在东亚古陆,是迄今已知古近纪较老而繁盛的一个昆虫群,具有地域性、时代性、环境性和过渡性等鲜明特点,在演化关系上彰显出中生代/新生代之交、古近纪/新近纪之交昆虫群之间过渡及转折的特点;也是在亚热带气候条件下发展起来的森林沼泽至湖沼环境中形成的一个富有多样性的昆虫群,对研究昆虫演化、生物区系划分、环境分析和成煤时代的确定等具有重要意义。

洪友崇(1929—)

洪友崇,广东南澳人,古昆虫学家,北京自然博物馆研究员。1953年毕业于北京地质学院,1958—1960年留学苏联科学院,回国后先后在地质科学院及北京自然博物馆工作。他首次研究抚顺古昆虫群,为我国中生代、新生代昆虫研究作出重要贡献。

张立君(1935—)

张立君,河北盐山人,古无脊椎动物学家,沈阳地质矿产研究所研究员,沈阳师范大学教授,专长于介形类化石及中生代、新生代地层研究,他为辽宁及我国东北中生代介形类及生物地层研究作出重要贡献。

狐假虎威——雄蝎蛉

雄性蝎蛉(Panorpa sp.)的腹部末端具有一个膨大的,像刺一样的器官,并且总是向身体的上方弯曲,看上去就像蝎子的刺一样,实际上这只是蝎蛉的雄性生殖器官,并非攻击性武器。雄蝎蛉通过模仿凶猛的蝎子,制造一个马上发动攻击的假象,用以震慑潜在的捕食者。此外,雄蝎蛉一方面对其他动物"狐假虎威",另一方面对雌蝎蛉耍小聪明,为引诱雌蝎蛉同它交配,往往事先准备好精美的礼物(小型昆虫或节肢动物)去打动雌蝎蛉,更甚者,雄蝎蛉还模仿雌虫的求偶行为,诱使其他雄蝎蛉误呈上准备好的"定情礼物",既节省了筹备礼物的时间,又降低了自己被捕食的危险。化石蝎蛉(Hemerobiidae)产于辽宁北票,距今约1.25亿年。

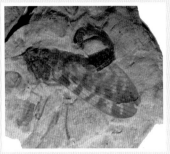

现生蝎蛉　　化石蝎蛉

抚顺生物群的其他门类化石主要包括叶肢介、介形类、腹足类、蝎类、鱼类、爬行类和藻类等。叶肢介化石也很丰富,主要产于西露天组,以抚顺雕饰叶肢介(*Fushunograpta*)组合带为代表,时代为始新世早期;由于该属在我国南方和西北一些省区广泛分布,时代多为始新世,因此,该组合也具有重要生物地层学意义。该组合之下为古新世近渔乡叶肢介(*Perilimnadia*)组合,之上是始新世古狭叶肢介(*Paraleptestheria*)组合。这三个叶肢介组合的属种都很单调,似表明第三纪时期叶肢介的演化已处于衰落。抚顺生物群的介形类化石类型和个体数量均较多,但保存较差,主要产于西露天组,偶见于计军屯组和耿家街组,以新美星介—美丽金星介(*Cyprinotus novellus-Cypris bella*)组合为代表,已发现13属27种。腹足类化石主要产于西露天组,以豆螺科类群为主,如弯口豆螺(*Bithynia ardostoma*)等。蝎类主要发现于古城子组琥珀中,如辽宁拟蝎(*Trachychelifer liaoningense*)等。鱼类化石较少,多为鱼鳞和鱼牙化石,主要产于计军屯组,其中偶见个体较大的鲤科成员。龟化石数量较少,仅在计军屯组油页岩中见满洲无盾龟(*Anosteira manchuriana*)等。

古近纪抚顺生物群所包括的微体生物群部分主要见于辽河油田附近的渤海湾盆地。微体生物化石异常繁盛。渤海湾盆地始新世—渐新世生物群曾被称为"华花介动物群"、"渤海湾生物群"、"华花介—渤海藻生物群"、"华花介—渤海藻科生物群"或"真星介—异星介—宽轮藻生物群"等(张立君,2011*)。本书建议统称为"渤海湾生物群"或"抚顺生物群渤海湾部分",其赋存层位主要是古近系的孔店组(房身泡组)、沙河街组和东营组。古近纪渤海湾的生物化石主要由介形类、腹足类、双壳类、有孔虫、多毛类、海绵、棘皮动物、

---

* 张立君. 2011. 辽宁古近纪抚顺生物群和渤海湾生物群(未刊)

鱼类、哺乳动物、昆虫、孢粉、藻类、疑源类和轮藻等门类组成。这一生物群是在偶受海泛影响的近海盆地中发生、发展和消亡的。湖盆水体深浅、含盐度和古地理变化等对水生生物的繁衍与分布起着重要控制作用,因而它是在淡水—半咸水环境交替及湖边陆地生存的生物共同构成的生物群。

渤海湾地区的介形类物化石迄今已发现41属500种以上,自早至晚可划分为:(1)五图真星介(*Eucypris wutuensis*)组合;(2)中国华北介(*Huabeinia chinensis*)组合;(3)椭圆拱星介(*Camarocypris elliptica*)组合;(4)惠民小豆介(*Phacocypris huiminensis*)组合;(5)单峰华花介(*Chinocythere unicuspidata*)组合;(6)弯脊东营介(*Dongyingia inflexicostata*)组合等。

渤海湾地区古近纪的孢粉化石十分丰富,主要产于始新统—渐新统孔店组、沙河街组和东营组,被划分为3个发展阶段:(1)杉科繁盛阶段(孔店组二段至沙河街组四段);(2)栎粉属繁盛阶段(沙河街组三段至一段);(3)榆粉属繁盛阶段(东营组)。孢粉化石的组成被划分为3个组合、9个亚组合(本书从略)。与此同时,该地区古近纪沟鞭藻类和疑源类也十分丰富,保存较完好。

1.2.10 辽宁的古人类与第四纪哺乳动物群

据古人类学研究,人类进化大致经过直立人(*Homo erectus*)、早期智人(early *Homo sapiens*)和晚期智人(late *Homo sapiens*)三个阶段。以往人类学家常称为"猿人"、"古人"及"新人"三个阶段。古人类与猿的区别主要在于身体已经能直立行走、脑容量的增加以及能制造工具。早期人类多以天然洞穴为居所,以狩猎采集为主要生活手段,能制造简

大连"金远洞"第四纪哺乳动物化石点

专家考察"金远洞"化石产地(2015.10)
左起:刘金远,傅仁义,金昌柱,孙革,赵博,高春玲。

"金远洞"哺乳动物化石

陋、粗糙的打制石器,会使用火和控制火种。

目前,国际上人类学家普遍认为,人类起源于大约300多万年前的非洲或亚洲南部。我国发现最早的为云南元谋人(距今170万年),其次为:陕西兰田人(距今90万—80万年)、周口店北京人(距今60万—50万年)以及辽宁庙后山人(距今50万—45万年)。辽宁地区地处辽河流域,是古人类生存、繁衍和发展的重要地区之一,至少约50万年前已有直立人阶段的本溪"庙后山人"活动的足迹。2014年,大连自然博物馆与中国科学院古脊椎动物与古人类研究所又在大连复州湾附近的驼骆山"金远洞"发现距今约70万年的第四纪哺乳动物新的化石点。

辽宁的古人类化石的发现始于20世纪50年代。1956年首次报道了建平县南地乡出土的"建平人"上臂骨(右侧肱骨)化石的发现,伴生有披毛犀、转角羚羊、野牛、野马等化石,其时代为晚更新世晚期(距今约1.5万年),体质特征与晚期智人一致[39]。20世纪70年代后期起,辽宁省古人类研究取得空前进展,首次发现"庙后山古人类化石群"[15,20]、"金牛山古人类化石群"以及其他各阶段的古人类化石和旧石器文化遗址20余处,代表了旧石器时代早、中、晚三个时期。由此,辽宁已初步建立起全省古人类及旧石器文化发展序列,为人类起源和我国古人类演化发展等研究提供了重要资料。与古人类相伴生的动物化石主要为第四纪哺乳动物化石群,以猛犸象—披毛犀动物组合为代表,迄今已发现化石点80余处。

1.2.10.1 庙后山古人类

庙后山古人类化石发现于本溪县山城子村采石场,洞穴高度相当于洞穴前的汤河的三级阶地,山体基岩由奥陶系马家沟组厚层灰岩组成。1978—1982年先后进行过3次发掘,洞穴内发现大量第四纪哺乳动物化石以及人类打制石器和骨器、人类用火遗迹、古人类牙齿化石。古人类化石包括1枚犬齿、右侧上白齿、下白齿和一段股骨化石。其中右侧上犬齿属老年人个体。从体质形态特征和出土层位观察,定为直立人阶段,其他几枚牙齿

贾兰坡院士(左2)考察庙后山遗址
(1984;右2为傅仁义教授)

庙后山古人类遗址

智人第一下臼齿

直立人犬齿

庙后山古人类化石

傅仁义教授在庙后山遗址
向国内外专家作介绍(2007)

李廷栋院士(右5)等参观庙后山古人类遗址(2008)

高星教授(右)等在庙后山古人类遗
址发掘(2012)

属于早期智人。第四纪哺乳动物包括肿骨鹿和三门马等76种动物化石。1986年辽宁省博物馆及本溪市博物馆合作发表了《辽宁本溪市旧石器文化遗址》[20]，2013年黄立强发表《庙后山——东北人最早的家园》[15]，1984年我国著名古人类学家贾兰坡院士曾亲临庙后山遗址考察，2012年中科院古脊椎动物与古人类研究所科考队对庙后山古人类遗址进行了新一轮发掘与研究。

据堆积地层同位素年代测定、古地磁和碳酸盐铀系法测定，庙后山人类化石的时代，为距今50万—45万年。庙后山人类化石的发现对研究早期人类分布、人类体质特征形态和人类迁移发展等，具有重要学术价值。

1.2.10.2　金牛山古人类

金牛山古人类化石点位于营口大石桥市西南8千米的永安镇西田村一山丘，面积约0.3

我国东北人类第一缕炊烟升起的地方——庙后山

庙后山遗址

庙后山位于辽宁省本溪县山城子村南的山坡上，是我国东北迄今最早的古人类化石和旧石器早期遗址。1978—1980年傅仁义等辽宁省考古专家在这里进行了三次发掘，发现了古人类牙齿及骨骼化石以及肿骨鹿和三门马等76种动物化石，与此同时还发现部分石器、骨器以及炭粒、灰烬等遗迹。1984年我国著名古人类学家贾兰坡院士曾来庙后山考察；2007年本溪市举行国际地层古生物学术研讨会，掀起了庙后山古人类研究的新热潮。2012年中科院古脊椎动物与古人类研究所科考队在庙后山开展了新一轮较大型化石发掘与研究，又新发现一批食虫类啮齿动物化石及动物化石碎片，还发现百余件石器、骨器及烧焦炭化的动物碎骨化石等。

庙后山古人类遗址的发现表明，早在距今约50万年前，与北京山顶洞人、周口店人在华北生活的同时，辽东地区已经有人类生存活动；而且，庙后山人已经掌握了用火、敲骨吸髓的食用方法等生存本领。古人类学专家傅仁义认为，庙后山周边阶地发育较好，有山有河，适合奇蹄类、偶蹄类动物生存，而古人类在1万年以前也基本都靠采集和狩猎为生；庙后山有大量动物存在作为食物来源，当然也适合古人类生存。众多的化石材料和地层证据表明，庙后山地区的古人类包括了直立人、早期智人和晚期智人3个时期的类群，距今50万年以来都有人类活动；这里山洞朝阳，洞前有阶地，距河流近，取水简便，适于古人类在这里繁衍生息。此外，对庙后山的古人类文化研究还表明，这里早期人类文化和华北的旧石器文化有着密切的联系；这对于探索中国远古人类文化的起源与发展，具有十分重要的意义。

庙后山古人类遗址的发现，填补了东北地区早期人类历史研究的空白；庙后山也被称为我国"东北第一人的故乡"和"我国东北人类第一缕炊烟升起的地方"。

  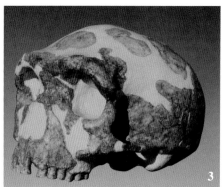

金牛山古人类遗址
1. 洞穴；2、3. 金牛山人头骨。

平方米，山体基岩由晚元古宙大石桥组白云质灰岩及泥质板岩等组成。金牛山A点洞穴最为典型，洞内堆满第四纪堆积物，发现了丰富的古人类化石、旧石器文化遗物及大量哺乳动物化石。铀系法测定距今28万年，相当中更新世晚期，文化时代为旧石器时代早期。

金牛山人化石发现于1984年，化石包括较完整的头骨（未见下颌骨），5个脊椎骨、2根肋骨，以及尺骨、髌骨、腓骨、腕骨、指骨等共55件，同属一个个体。根据人体结构特征等分析，应为青年女性个体，处于直立人向早期智人过渡阶段。时代为中更新世。金牛山人的发现是继周口店北京猿人之后，我国北方旧石器时代考古又一重大发现。1984年被评为我国十大考古发现之一，被联合国科教文组织评为世界十大科技进展之一。金牛山古人类遗址1998年被国务院颁布为全国重点文物保护单位。

1.2.10.3 辽宁其他古人类化石点

辽宁其他重要古人类化石还包括有：**建平人**（建平县南地乡，晚更新世晚期，距今约1.5万年），**鸽子洞人**（喀左县水泉乡瓦房村大凌河畔，晚更新世中晚期，距今7万—5万年，旧石器时代中期遗址），**小孤山人**（海城市孤山镇孤山村青云山下，晚更新世晚期，距今约4万年，相当于旧石器时代中晚期），**丹东前阳人**（丹东东港市前阳镇白家堡采石场，时代距今约1.86万年，晚更新世晚期，相当于旧石器时代晚期）等。

1.2.10.4 辽宁第四纪哺乳动物化石群

与辽宁古人类化石相伴发现的为第四纪哺乳动物化石群，化石也十分丰富，迄今在辽宁共发现80余处，主要分布于辽宁丘陵地带的石灰岩洞穴以及辽河流域河漫滩阶地或坡积上的黄土堆积中，时代囊括了第四纪各阶段，包括更新世早期（距今300万—100万年），更新世中期（距今100万—20万年）以及更新世晚期（距今20万—1万年）等。典型的哺乳

辽宁其他古人类化石点
1. 建平人右肱骨；2. 鸽子洞人化石产地(喀左县水泉)；3. 海城小孤山人产地；
4. 丹东前阳人化石产地；5. 丹东前阳人头骨；6. 丹东前阳人下颌骨。

动物化石群主要包括大连海茂动物群、本溪庙后山动物群、营口金牛山动物群、喀左帽儿山动物群、辽阳安平动物群、营口藏山动物群、喀左鸽子洞动物群、朝阳马山洞动物群等。

大连海茂动物群

位于大连市甘井子区海茂村采石场。动物化石包括食虫目、翼手目、啮齿目、兔形目(刺猬、小鼹鼠、更新菊头蝠、泥河湾短耳兔、原鼢鼠、原东北鼢鼠、五趾跳鼠、小根田鼠)等共4个目11科24属28种。该动物群以小型哺乳动物为主，大部分为原始的新近纪残余种，反映干旱草原生态类型，时代为早更新世，距今160万—120万年。这是迄今我国东北地区最早的哺乳动物化石点。

本溪庙后山动物群

位于本溪市本溪县山城子乡庙后山采石场洞穴，位置同庙后山人。动物化石主要包括喜马拉雅旱獭、达呼尔鼠兔、中华貉、普氏野马、河套大角鹿、普氏羚羊、安氏中华河狸、师氏中华河狸、三门马、梅氏犀、中国缟鬣狗、杨氏虎、似剑齿虎、肿骨鹿、李氏野猪、硕猕猴等。基本上是华北中更新世典型动物种属，也有少量早更新世甚至新近纪残余种，与周口店第1地点中下部相似。地质时代为中更新世中晚期，距今约40万—23万年。

营口金牛山动物群

位于营口大石桥市南8千米的永安乡西田屯村，位置同金牛山人。主要动物化石有：

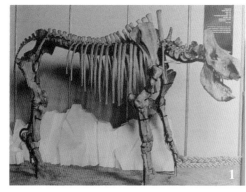
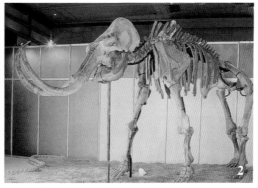

辽宁第四纪猛犸象—披毛犀动物群化石
1. 披毛犀；2. 猛犸象；3. 猛犸象复原。

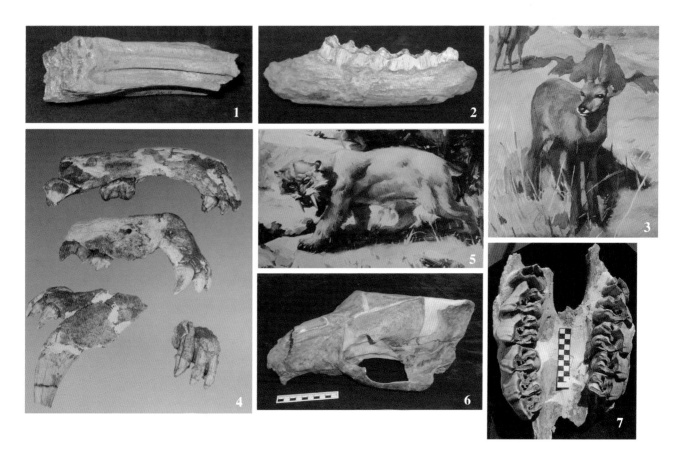

辽宁第四纪哺乳动物群化石（Q_2）
1. 三门马（牙齿）；2、3. 肿骨鹿及复原；4、5. 剑齿虎（上牙床）及复原；6. 棕熊头骨；
7. 梅氏犀。1、2产自庙后山，Q_2；4、6、7. 产自金牛山，Q_2。（据辽宁古生物博物馆）

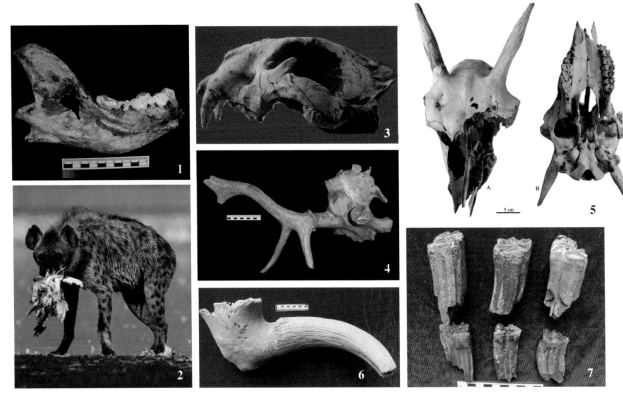

辽宁第四纪哺乳动物化石（Q_3）

1,2. 斑鬣狗及复原；3. 虎；4. 马鹿；5. 羚羊；6. 东北野牛；7. 普氏野马。

1—5 产自马山洞，Q_3；6、7 产自昌图曲家店，Q_3。

蒙古黄鼠、黑线仓鼠、南鼬、恰克图转角羚羊、韩氏刺猬、硕猕猴、翁氏兔、东北鼠兔、变异仓鼠、古田鼠、居氏大河狸、拉氏豪猪、变种狼、中国貉、最后斑鬣狗、最后似剑齿虎、鬃猎豹、三门马、梅氏犀、肿骨鹿、赤鹿、东北马鹿、北京香麝、李氏野猪等，共 8 个目 50 余种，此外还发现爬行类龟甲、鱼类、鸟类、鸽子、鸭、金雕、环颈雉、驼鸟等。时代为中更新世晚期至晚更新世早期，距今 31 万—16 万年。该化石动物群的规模目前居我国东北第四纪动物群之首。

喀左帽儿山动物群

位于喀左县兴隆庄乡章京营子村东山嘴采石场。主要化石类群包括：鼢鼠、翁氏兔、鬣狗、熊、三门马、野猪、肿骨鹿等，是迄今辽西地区最早的新生代化石地点，动物群面貌与北京周口店动物群相似，时代为中更新世中晚期。

辽阳安平动物群

位于辽阳市弓长岭区南山采石场。主要化石类群包括：田鼠、方氏鼢鼠、硕猕猴、杨氏虎、沙狐、中国貉、梅氏犀、葛氏斑鹿、水鹿、北京麝、野猪等共 19 个属种，以食肉目和偶蹄目为主，梅氏犀数量最多，时代为中更新世中晚期。

营口藏山动物群

位于营口大石桥市百寨子乡陈家村以北2.5千米采石场。动物化石主要有：布氏田鼠、巢鼠、硕猕猴、变异狼、貉、西伯利亚鼬、狗獾、最后斑鬣狗、三门马、梅氏犀、转角羚羊等，共6目13科18属21种；此外有瓣鳃类、腹足类、龟甲片和鸟类等化石。该动物群主要为喜温暖环境的动物，面貌与周口店顶部堆积的相似，时代为中更新世晚期，距今20万年左右。

喀左鸽子洞动物群

位于喀左县水泉乡瓦房村大凌河畔，位置与鸽子洞人相同。主要化石有达呼尔鼠兔、硕旱獭、沙狐、中华猫、最后斑鬣狗、棕熊、野马、葛氏斑鹿、普氏羚羊、岩羊等26个属种。该动物群有华北更新世动物群成员，也有东北晚更新世猛犸象—披毛犀动物群成员，其绝灭种占全部属种的30.7%，绝大多数类群代表寒冷环境生存的动物。时代为晚更新世中晚期，距今7万—5万年。

朝阳马山洞动物群

位于朝阳市龙城区边杖子乡朱杖子村马山采石场。主要动物化石有獾、狼、虎、披毛犀、转角羚羊、普氏羚羊等，时代为晚更新世中晚期，距今约10万年。这批动物化石以偶蹄目和奇蹄目等草食性动物为主，数量极其丰富，骨架保存完整，在辽宁乃至东北地区罕见，对研究新生代哺乳动物的演化和生态特征等具有重要价值。

其他晚更新世动物群

辽宁其他的晚更新世动物群还包括有：辽阳三星堆，昌图八面城，本溪三道岗及本溪湖，铁岭李千户，大连古龙山、金州湾里及旅顺口黄海海底，海城小孤山，喀左二布尺，凌源西八间房，丹东前阳、通远堡，开原粮窖村，东港黄土坎，东鞍山下房身、大井村，法库丁家房及叶茂台，彰武大冷乡，满堂红闹德海水库，新民东蛇山，锦县沈家台，以及建平人化石点等地所发现的动物群或化石点等。

第二章

辽宁古生物博物馆巡礼

　　辽宁是我国乃至全世界的"化石宝库",她的热河生物群及燕辽生物群等珍贵化石为研究全球生命起源与演化作出了重要贡献,被誉为地球上"第一只鸟起飞、第一朵花绽放的地方"。"鞍山群早期生命"化石等将辽宁的地质历史记录向前推进至距今30亿—25亿年,使辽宁成为我国具有最古老的地质记录的省份之一。

　　为进一步保护和研究辽宁特有的化石资源,2006年经辽宁省人民政府批准,由辽宁省

辽宁古生物博物馆外景(正面)

国土资源厅与沈阳师范大学合作建设"辽宁古生物博物馆"。经过五年的辛勤建设,这座我国迄今规模最大的古生物博物馆于2011年5月21日正式对外免费开放。

辽宁古生物博物馆(Paleontological Museum of Liaoning,简称PMOL)位于沈阳师范大学主校门北侧,西临黄河北大街,地处沈北"大学城"的起点。该馆占地面积19 000平方米,建筑面积15 000平方米,建筑外形像一个庞大的地质体和一个巨型恐龙巧妙的融合:断层将地质体垂直切割,火山熔岩自上而下奔泻流淌,带观众走进辽宁30多亿年地质历史的长河。南侧的拱形建筑代表辽宁巨大恐龙的身躯,中间的钢架代表恐龙脊柱,两侧是恐龙的肋骨,而球体是恐龙蛋。21根恐龙肋骨的钢架象征辽宁人民21世纪挺拔的身躯,预示着辽宁美好的前程。

辽宁古生物博物馆以科学性为主,以展示地史时期生命起源与演化为主线,以介绍30亿年来辽宁"十大古生物群"为重点,具有鲜明的国际化特色。馆内共设8个展厅16个展区。第3厅介绍30亿年来辽宁"十大古生物群",其中,鞍山群早期生命、燕辽生物群、热河生物群、辽宁的古人类是展区中的四大亮点。第4厅介绍世界著名的热河生物群,带观众

辽宁古生物博物馆外景(侧面)

走进1亿多年前辽西的"恐龙王国"、"古鸟世界"、"花的摇篮"等奇妙的史前世界。第8厅是气势磅礴的"辽宁大型恐龙厅",8件产自辽宁的大型恐龙首次集体"亮相",其中长达15米的辽宁巨龙由辽宁古生物博物馆首次发现和研究,也是第一次与观众见面。

辽宁古生物博物馆的建立是辽宁省文化教育事业和我国古生物学事业的一件大事,对提升辽宁省的科学普及水平、加强辽宁省古生物化石的保护与研究,以及促进辽宁省的国际交流与文化旅游事业发展等,发挥着重要作用。该馆的建立对中国古生物学事业和博物馆事业的发展也是一个"功在千秋"的壮举。2012年9月,时任国土资源部部长、现任国家发改委主任徐绍史在参观博物馆后赞扬说,辽宁古生物博物馆是政府与高校合作建设博物馆的一个成功范例。

2.1 博物馆设计与展陈

辽宁古生物博物馆建筑面积15 000平方米,包括地下部分1000平方米,地上部分约14 000平方米;北侧主体建筑最高点29.2米;共分5层(地上4层,地下1层),地上层高约6.0米;建筑的北侧为钢筋混凝土框架结构,南侧为钢排架结构。

2.1.1 博物馆的外形设计

辽宁古生物博物馆的建筑外形犹如精美的艺术品:庞大的地质体(主体)和一只巨型恐龙(侧体)巧妙地融合,寓示着挺拔的恐龙产自体魄巨大的母体——中生代地层之中;中间的分离源自断层,火山熔岩沿断层两侧倾泻而下,仿佛回到1亿多年前辽宁大地上生命与大自然竞相争雄、又相互依存的地质景观。辽宁古生物博物馆的建筑宛如一个巨

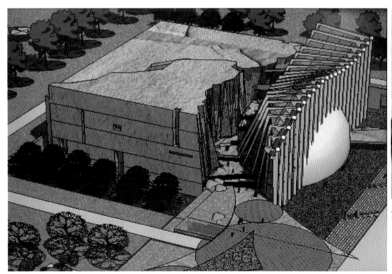

辽宁古生物博物馆设计师李祖原

辽宁古生物博物馆外形设计模型

型艺术雕塑,给观众留下的"第一印象"是雄伟、壮观。该建筑外形设计由著名建筑设计大师李祖原先生领导的建筑设计事务所完成。建筑施工设计及其实施由辽宁省建筑设计院完成。

李祖原先生是国际著名建筑师,毕业于台湾成功大学建筑系,于美国普林斯顿大学获建筑学硕士学位。他一直致力于研究和创造有中国传统特色的新建筑,以"生命建筑"为设计理念,结合现代科技,提供高创意性、高整合性、高技术性的建筑设计服务。他曾主持设计台北101大楼。此次辽宁古生物博物馆的建筑外形设计,又成为他和他的设计团队的绝世佳作之一。

2.1.2 博物馆的展厅

走进辽宁古生物博物馆,整个展览都贯穿着"生命进化"的主线。展览以介绍"30亿年来辽宁十大古生物群"为重点,突出展示"热河生物群"、"燕辽生物群"、"鞍山群早期生命"及"辽宁古人类"等四大亮点,集古生物展示、收藏、科研、科普与教学五大功能为一体。

辽宁古生物博物馆展示(区)主要包括:

第1厅　辽宁古生物博物馆简介(包括序厅"地质长廊")
展区Ⅰ　辽宁古生物博物馆简介

第2厅　地球与早期生命
展区Ⅱ　地球及其生命起源与早期演化
展区Ⅲ　寒武纪生命大爆发

第3厅　30亿年来的辽宁古生物
展区Ⅳ　辽宁地质历史概况
展区Ⅴ　辽宁十大古生物群

第4厅　热河生物群
展区Ⅵ　热河生物群简介
展区Ⅶ　恐龙王国
展区Ⅷ　古鸟世界
展区Ⅸ　花的摇篮

展区 X　其他生物化石

第 5 厅　国际古生物学与世界各地化石
展区 XI　古生物学发展史与古生物学家
展区 XII　世界各地的古生物化石
展区 XIII　恐龙大灭绝与 K–Pg 界线
展区 XIV　国际古生物学交流与合作在辽宁
附:5-1　中央恐龙区

第 6 厅　互动科普
附:3D 影院

第 7 厅　珍品化石
展区 XV　珍品化石

第 8 厅　辽宁大型恐龙
展区 XVI　辽宁大型恐龙

在第 1 厅"辽宁古生物博物馆简介"展厅,以辽宁 30 亿年来地质演化史为实例的"地质长廊"首先进入眼帘。从五颜六色岩石组成的地质剖面中,观众可以大致了解太古宙、元古宙、古生代、中生代和新生代等各时期整个中国北方地质古生物环境变迁,地质剖面对应着古环境背景,使观众对"燕辽生物群"、"热河生物群"、"辽宁古人类"以及"辽宁 4 大主要成煤期"等有初步印象。

"地质长廊"与"地球与早期生命"展示　1,2. 地质长廊(第 1 厅);3. 地球与早期生命(第 2 厅)。

走进第2厅"地球与早期生命"展厅,主要感受自46亿年前地球出现以来,早期生命出现及演化历程的漫长和艰辛。其中包括介绍地球形成初期(46亿—38亿年前)原始海洋从无机阶段到有机阶段的演化,辅之以著名的"米勒实验",以及原始生命在长达近30亿年的演化中出现早期后生动物(包括伊迪卡拉型生物群等)的曲折进程等。当然,从云南澄江"寒武纪生命大爆发"、贵州"凯里生物群"和加拿大"布尔吉斯生物群"的众多化石,以及产自大连海滨的叠层石等,会使观众真正看到生命演化的曙光。

产自大连的叠层石

第3厅"30亿年来的辽宁古生物"和第4厅"热河生物群",是辽宁古生物博物馆最具特色的两个展厅。第3厅重点介绍了辽宁地质历史概况和30亿年来的辽宁"十大古生物群";其中,太古宙—古生代包括鞍山群早期生命、寒武纪—奥陶纪海洋生物群、石炭纪本溪生物群等3个生物群,中生代包括中三叠世林家生物群、晚三叠世羊草沟生物群、燕辽生物群、热河生物群和阜新生物群等5个生物群,新生代包括抚顺生物群和辽宁的古人类及第四纪哺乳动物群等。为凸显辽宁的古生物特色,展陈中将第4厅作为著名的热河生物群的专题展厅,以"恐龙王国"、"古鸟世界"、"花的摇篮"和"伴生生物"等4个板块和众多精美的化石,详细展示了热河生物群的面貌及其

辽宁十大古生物群
1. 第3厅的"鞍山群早期生命"展台;2,3. 第4厅的"恐龙王国"展台。

在生物演化研究中的意义。

第5厅"国际古生物学与世界各地化石"展厅主要展出来自德、英、俄、美、澳、日、印、泰、阿富汗及爱沙尼亚等10多个国家赠送的古生物化石，其中以德国森肯堡自然博物馆赠送的24件麦索化石（复制品）最为精彩。

在二楼中央大厅，即第6厅"互动科普"，观众可以进行与博物馆之间的互动，包括参加"与恐龙赛跑"、"与恐龙照相"、"催恐龙下蛋"、参观"恐龙剧场"以及在3D影院观看《会飞的恐龙》等科普影片。

第7厅"珍品化石"展厅展出的是辽宁古生物博物馆的"镇馆之宝"，包括赫氏近鸟龙、巨齿兽、赵氏翔龙、沈师鸟、大平房鸟、辽龟、辽宁古果、中华古果以及39个幼年个体在一起的"鹦鹉嘴龙幼儿园"等世界级化石珍品，让观众一饱眼福并惊赞它们的科学意义。第8厅"辽宁大型恐龙"展厅首次展示了以辽宁巨龙为代表的、产于辽宁白垩纪的8件大型恐龙，令观众震撼并流连忘返。

第5厅德国麦索古生物化石

第7厅珍品化石

第6厅科普互动：恐龙生蛋（左）、与恐龙赛跑（右）

2.1.3 展陈设计理念

辽宁古生物博物馆内部展陈的总体设计是由我国古生物学家、辽宁古生物博物馆馆长孙革教授带领博物馆的专家们亲自完成的。展陈设计的总体理念一是科学性,二是国际化,三是突出辽宁。设计中还充分考虑了展陈要结合辽宁的化石保护工作的宣传、高科技手段的应用,以及重视科学家亲自参与及指导展陈设计等。

科学性理念主要是:展陈设计强调科学性,包括以生物进化为主线,高水平地介绍地质古生物知识,宣传达尔文的生命进化理论,注重化石及其相关展示的科学内涵和学术支撑等。本次展陈设计中,结合辽宁的化石,突出介绍了生命的起源与早期演化、带毛恐龙在研究鸟类起源中的意义、辽宁发现的迄今"最早的花"在研究全球有花植物起源中的作用,以及辽宁古人类化石发现在我国及东北亚人类演化与迁徙研究中的重要意义等。展览中,辅以科学著作、科学实验实例及古生态景观的最新复原的介绍等,例如,赫氏近鸟龙、赵氏翔龙、迄今世界"最早的花"等标本展示均有国际权威学术刊物[《科学》、《自然》及《美国科学院院刊》(*PNAS*)等]的论文随同介绍等,使观众对展品的科学内涵有更深刻的了解和领会。

国际化理念主要是:学习和借鉴当今世界主要优秀的自然类博物馆(如英国伦敦自然史博物馆、美国史密森自然史博物馆及纽约自然史博物馆、德国森肯堡自然史博物馆,以及日本福井县立恐龙博物馆等)的经验,将国际先进的设计理念融入辽宁古生物博物馆展陈中。与此同时,为加强国际交流与合作以及激发国内外观众的兴趣,展品中增加了世界各地化石标本及相关实物,国际自然科学家、博物馆学家的生平、成就、主要学说等,以及辽宁古生物博物馆国际合作与交流图片及实物等,用以扩大观众的国际古生物学视野,并扩大国际交流。如第5厅("国际古生物学与世界各地化石"展厅)中有美国、德国、英国、法国、俄罗斯、日本、澳大利亚、泰国、阿富汗、印度、加拿大、南非及爱沙尼亚等13个国家

第8厅"辽宁大型恐龙"

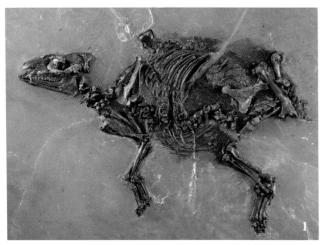

德国麦索化石展区设计
1. 德国麦索小始祖马（*Hyracotherium*）复制品；
2. 德国森肯堡博物馆专家来馆安排麦索化石展览，设计师容凯拉（E. Junqueira, 中）及沙尔（S. Schaal）教授（左3）（2011.3）。

的古生物化石（包括复制品）展品数十件，包括德国森肯堡自然史博物馆惠赠的德国始新世麦索化石及德国始祖鸟（*Archaeopteryx*）复制品等，这些化石复制品都是首次集中与我国观众见面。

突出辽宁的理念主要是：以介绍"30亿年来辽宁十大古生物群"为重点，以生命进化为主线；以突出"热河生物群"、"燕辽生物群"、"鞍山群早期生命"及"辽宁古人类"等为特色。展陈中设计了展示赫氏近鸟龙、赵氏翔龙、辽宁古果及沈师鸟等"四大明星"。与此同时，设计突显全馆展陈的化石展品主要来自辽宁，特别是第8厅（"辽宁大型恐龙"展厅）展示的8具大型恐龙，包括辽宁巨龙、东北巨龙、北票龙、双庙龙、克氏龙、薄氏龙、朝阳禽龙及暴龙等，全为近年来古生物专家在辽宁采集与研究，其中，辽宁巨龙为辽宁古生物博物馆专家亲自采集并开展研究工作的。

此外，此次展陈设计还注重结合辽宁化石保护工作的宣传、高科技手段的应用、科学普及，以及注重科学家亲自参与及指导展陈的设计与实施。展陈设计中特别注意宣传了辽宁省在古生物化石保护工作中取得的成绩，以增强广大群众保护化石的责任感及法律意识。高科技手段方面，主要包括：（1）采用国际先进的和最新的多媒体化石生态展示视频与解说（如：上海睿宏公司的《会飞的恐龙》、中国CCTV的《破解讨厌之谜》、日本的《花的起源》及法国的动漫等）；（2）引进日本福井县立恐龙博物馆（本馆装配制作）的"恐龙剧场"；（3）北京天图公司的"与恐龙赛跑"和常州卓谨公司的"与恐龙合影"等互动方式。上述现代化展示手段的运用增强了观众特别是广大青少年对"生物进化论"知识的理解，在科普中发挥了重要作用。

辽宁古生物博物馆高度重视科学家亲自参与及指导展陈设计和施工。先后邀请了

辽宁古生物博物馆的"四大化石明星" 　　　　　　　　讲解员王璐介绍赫氏近鸟龙
左起：近鸟龙，沈师鸟，翔龙，古果。

我国著名恐龙学家董枝明、徐星，古鸟类学家侯连海等参与指导，大连自然博物馆的专家刘金远和高春玲、甘肃地质博物馆恐龙专家李大庆等都曾亲自来现场指导恐龙安装等工作。

2009年2月21日，由辽宁省国土资源厅和沈阳师范大学共同主办、对由孙革主持完成的《辽宁古生物博物馆展陈设计大纲》的评审会议在沈阳召开。评审委员云集了我国当今著名古生物学家与相关博物馆及化石管理等领域的专家，邀请我国著名地质学家刘嘉麒院士为评审委员会主任，评审委员包括董枝明、季强、徐星、高克勤、金昌柱、孙春林、王伟铭、任东、吕君昌、段吉业、郑少林、王五力、张立君、彭光照、续颜等，也包括辽宁省内外有关博物馆及化石管理方面的专家崔滨、孙永山、赵义宾、王丽霞等。沈阳师范大学党委书记于文明、前校长赵大宇和辽宁省国土资源厅前副厅长张殿双亲自到会，对展陈设计工作的评审高度重视。经评审，专家一致认为，《辽宁古生物博物馆展陈设计大纲》总体上达到了自然类博物馆展陈设计大纲领域的国际先进水平。

 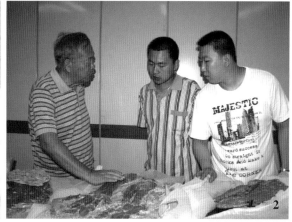

专家参与及指导展陈设计
1.董枝明(左)、徐星(中)在恐龙厅指导展陈布置；2.董枝明教授(左)指导恐龙标本修复。

出席辽宁古生物博物馆内部展陈设计大纲评审会的专家与领导

前排（左起）：金昌柱、孙革、季强、张殿双、赵大宇、刘嘉麒、于文明、董枝明、郑少林、王五力。后排（左起）：王丽霞、程绍利、彭光照、崔滨、吕君昌、高克勤、李崇春、徐星、张立君、段冶、孙春林、赵义宾、孙永山、李印先、王伟铭、胡东宇、段吉业、田明申、续颜、张立军、任东、周长付、刘玉双。

2.2 五年建设历程

辽宁古生物博物馆的建设，是辽宁省国土资源厅和沈阳师范大学在认真贯彻落实国家《古生物化石管理条例》工作中的一个成功实例，也是他们共同创造性地开展工作的一个具有"里程碑"意义的突出展示。该馆的建设既是国家古生物化石保护和研究的需要，也是国内外科学家和广大群众的众望所归。

2.2.1 一拍即合

1996年10月，国土资源部期刊《中国地质》发表了时任中国地质博物馆馆长季强研究员和姬书安研究员共同撰写的论文《中国最早鸟类化石的发现及鸟类的起源》[16]，首次报

辽宁省国土资源厅与沈阳师范大学商讨建设辽宁古生物博物馆（2005）

左图为辽宁省国土资源厅原厅长王大操（左）与沈阳师范大学前校长赵大宇（右）亲切会谈。

道了带毛恐龙"中华龙鸟"在我国的最早发现,一时间轰动国际学术界。1998年11月,国际权威学术刊物、美国《科学》杂志又发表了时任中国科学院南京地质古生物研究所孙革研究员等撰写的《追索第一朵花》,首次报道了迄今最早的被子植物"辽宁古果"在我国辽西的发现[85],进一步将全球的目光吸引到辽宁。随着一批孔子鸟、尾羽鸟(龙)、小盗龙等进一步发现,辽西的古生物化石不断展露出惊人的科学价值,它们在研究我国及全球生物演化中的作用越来越得到国际学术界的重视。从此,"辽宁是第一只鸟起飞、第一朵花绽放的地方"的赞誉传遍世界。

面对如此众多的重要化石不断被发现,辽西的古生物化石保护提上了主管部门——辽宁省国土资源厅的议事日程。2001年由辽宁省国土资源厅会同辽宁省法制办公室共同起草的《辽宁省古生物化石保护条例》经辽宁省人大通过,辽宁的化石保护已有法可依,但化石、尤其是一些重要化石的集中收藏已成为亟待考虑的问题。特别是,许多珍稀的古生物化石具有重要科学价值、急需研究,而辽宁自身还缺乏研究力量。因此,尽快在辽宁省省会沈阳建立省级古生物博物馆,便成为辽宁省人民政府、特别是辽宁省国土资源厅领导的共识。

为做好辽宁化石保护与研究工作,辽宁省国土资源厅领导自2002年起就开始筹划建设辽宁古生物博物馆。考虑到辽宁的古生物化石的研究与管理的科学性很强,辽宁省国土资源厅认为,博物馆最好由省国土资源厅与高校联合共建。但哪所大学能担此重任呢?时任辽宁省国土资源厅厅长王大操等领导通过多方调查了解,终于发现了"好伙伴"——沈阳师范大学,该校也愿意担此重任。于是,时任沈阳师范大学校长赵大宇和王大操厅长热诚会面商谈,双方"一拍即合",一致同

辽宁古生物博物馆开工典礼(2006.6.6)
1. 奠基;
2. 沈阳师范大学党委书记于文明致辞;
3. 省国土资源厅前厅长焉锦林宣布开工;
4. 与会嘉宾。

意由双方合作共建,辽宁古生物博物馆建在沈阳师范大学校园,由沈阳师范大学具体承担建设任务。

沈阳师范大学已有50余年历史,是一所设有哲学、经济学、法学、教育学、文学、理学、工学、管理学、艺术学等多个学科的综合性师范大学,在辽宁省的教育、科研及社会服务等方面一直作出重要贡献。该校领导工作力主创新,不断主动迎接新挑战。面对辽宁省的化石保护形势和科研工作需要,恰好沈阳师范大学又有生物学科研究人才,2005年学校适时成立了"辽西中生代古生物研究所"(后改名"古生物研究所"),聘请了我国著名古鸟类学家侯连海教授等专家前来指导,2007年起又聘请著名古生物学家孙革教授兼任所长,一批毕业于日本、澳大利亚的大学及北京大学等国内著名高校的年轻博士先后加盟。因此,沈阳师范大学已具有较为雄厚的人才实力,有能力承担此项重任。于是,在辽宁,一项由政府和高校共建古生物博物馆的工作就此开启。2006年6月6日,辽宁古生物博物馆正式动工兴建。

2.2.2 共筑大厦

在沈阳师范大学和辽宁省国土资源厅的共同努力下,经过近五年的风风雨雨,一座建筑面积15 000平方米、高约30米的巨型大厦——辽宁古生物博物馆拔地而起,填补了辽宁省省级古生物博物馆的空白。

五年的建设中,负责建筑施工的辽宁省建筑设计院精心施工,负责内部展陈施工的北京原动力公司(现天图公司)和沈阳瑞德公司以及负责外部辅助施工的鲁迅美术学院公司也都付出了辛勤的劳动。

博物馆建设历程
1. 前校长赵大宇(中)在建设工地;
2. 赵校长(右4)冒雨指导工作;
3. 季风岚厅长(中)视察博物馆施工;
4. 季厅长(左2)等视察展陈准备。

领导关怀

建设期间前来视察的部分领导。1. 前副省长鲁昕(右); 2. 前省委副书记陈希(左2); 3. 前副省长陈超英(右2); 4. 前国土资源部副部长蒋承菘(右); 5. 前国土资源部副部长寿嘉华(左3); 6. 前省发改委主任仲跻权(中)。

施工建设期间,沈阳师范大学前校长赵大宇、前副校长李崇春、后勤处长田明申等,辽宁省国土资源厅厅长季风岚、副厅长张殿双、杨德军,以及辽宁省化石管理局前局长孙永山等领导,都曾亲临现场指导工作。建设者们将建设"雏形"中的辽宁古生物博物馆视为心中的"宝贝",像雕塑家对待自己的雕塑作品一样,天天在精心打磨、细心爱护。

辽宁古生物博物馆的建设一直得到辽宁省委和省政府的关怀与支持,国家有关部委的领导和众多专家们也十分关心和支持。前副省长鲁昕、陈超英,前省委副书记陈希,前国土资源部副部长蒋承菘、寿嘉华,辽宁省发改委前主任仲跻权、辽宁省教育厅前厅长魏小鹏,以及我国著名地质学家李廷栋、刘嘉麒院士等领导和专家,都曾亲临沈阳师范大学视察指导古生物博物馆建设和筹备,并提出宝贵的意见和建议。

辽宁古生物博物馆建设的五年中,也得到了国际古生物学专家、博物馆领导以及众多国际友人的热诚关心和支持。美国植物学会前主席、科学院院士迪尔切,俄罗斯科学院通讯院士阿克米梯耶夫(M. Akhmetiev),德国科学院院士、森肯堡自然史博物馆馆长莫斯布鲁格(V. Mosbrugger),国际古植物学会主席、德国斯图加特自然史博物馆馆长怡德(J. Eder),全球最大的自然史博物馆、美国史密森研究院博物馆馆长约翰森(K. Johnson),国际

国际科学界与博物馆学界的支持
德国斯图加特博物馆建馆前提前签署合作协议。
前排左起：赵大宇（前校长），怡德馆长，
莫斯布鲁格馆长（院士），孙革馆长。

著名恐龙学家、比利时皇家自然史研究所教授哥德弗洛依特（P. Godefroit）等，都来过辽宁古生物博物馆建设工地。他们不仅称赞辽宁古生物博物馆的建设壮举，还提出了包括增加研究室和实验室建设面积等宝贵建议。在德国森肯堡自然史博物馆馆长莫斯布鲁格院士支持下，该馆无偿向辽宁古生物博物馆赠送了24件珍贵的麦索生物群化石标本复制品；斯图加特自然史博物馆馆长怡德还率先与建设中的辽宁古生物博物馆签署了合作协议。

因此，辽宁古生物博物馆的建设成功，不仅是众多设计者、施工者、管理者和科学家智慧与汗水的结晶，也是我国在自然类博物馆建设中得到国际友人支持、与国际学术界合作的见证。

2.2.3 开馆典礼

2011年5月21日，沈阳师范大学校园一片欢腾，辽宁古生物博物馆开幕庆典仪式在这里隆重举行。国家教育部副部长鲁昕、国土资源部副部长汪民、辽宁省前副省长陈超英，中国科学院院士李廷栋、刘嘉麒、周忠和，美国科学院院士迪尔切，以及来自美国、英国、德国、法国、俄罗斯、日本、蒙古、韩国、越南、印度、巴西

国外专家来博物馆建设工地参观

及我国的近百余名专家学者和来宾出席了庆典仪式，满怀喜悦地见证了辽宁古生物博物馆的开幕，一座我国迄今规模最大的古生物博物馆终于与国内外观众见面。

开馆典礼
1. 辽宁古生物博物馆大楼；
2. 沈阳师范大学党委书记于文明主持开幕式；
3—6. 领导及嘉宾致辞（3. 鲁昕副部长，4. 汪民副部长，5. 季风岚厅长，6. 迪尔切院士）；
7. 鲁昕副部长和汪民副部长等参观博物馆；
8. 出席典礼的国内外来宾。

出席开幕式的国内外来宾在恐龙大厅合影(2011.5.21)

2.3 十大化石明星

辽宁古生物博物馆开馆时,曾命名了赫氏近鸟龙、沈师鸟、赵氏翔龙及辽宁古果等4件化石珍品为馆内"四大明星"。近年来,随着化石采集与研究的深入,一批新的珍品化石不断涌现。为此,本书提出辽宁古生物博物馆"十大化石明星",用以对以往"四大明星"的补充。它们是:(1)赫氏近鸟龙,(2)辽宁巨龙,(3)鹦鹉嘴龙("幼儿园"标本),(4)巨齿兽,(5)翔龙,(6)古果,(7)孔子鸟,(8)大平房鸟,(9)辽龟,(10)沈师鸟。

2.3.1 近鸟龙(*Anchiornis* Xu,2009)——世界最早的带羽毛恐龙

赫氏近鸟龙(*Anchiornis huxleyi*)是胡东宇教授课题组2009年在英国《自然》杂志首次报道的迄今世界最早的带羽毛恐龙。近鸟龙是徐星教授2009年初命名的,是与鸟类亲缘关系最近的一种小型兽脚类恐龙。

赫氏近鸟龙的前、后肢和尾部均分布奇特的飞羽,其趾爪以外的趾骨上也都被有羽毛。胡东宇等提出,近鸟龙属于原始的伤齿龙类,兽脚类恐龙在晚侏罗世之前可能都已出

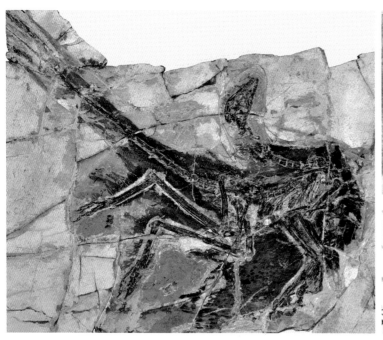
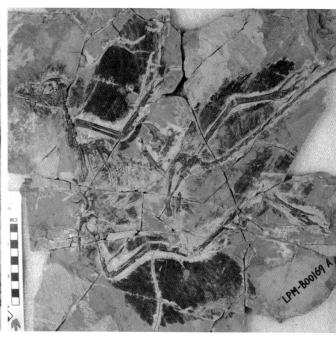

赫氏近鸟龙

现且迅速分化,包括鸟类在内的许多重要类群就是在这次快速演化事件中出现的。赫氏近鸟龙化石发现于建昌侏罗纪髫髻山组,距今约1.6亿年,较以往所知最早的鸟化石——德国始祖鸟早约1000万年,代表了目前世界上最早的长有羽毛的物种。它不仅进一步支持了恐龙演化曾经过"四翼阶段"的假说,也为鸟类起源于兽脚类恐龙的假说提供了强有力的证据。此成果代表了鸟类起源研究的一个新的、国际性的重大突破,曾入选"2009年中国高校十大科技进展"及"2009年世界/中国十大科技进展新闻"等。

产地与层位:辽宁建昌玲珑塔大西山;侏罗纪髫髻山组(距今约1.6亿年)。

2.3.2 辽宁巨龙(*Liaoningotitan*,MS*)——辽宁最大的恐龙

辽宁巨龙是巨龙形类恐龙的一个新的分类群,也是迄今辽宁发现的最大的恐龙。恐龙骨架保存近完整,体长约15米,前肢较短,约为后肢长的70%。头骨和下颌保存较完整,包括具凸形的口缘;颧骨位置靠前,接近眶前窗的前缘;上颌齿列短,位于口缘的前部;上颌牙齿呈迭瓦状,齿冠窄,断面呈D字形;下颌齿9枚,较小,未呈迭瓦状排列;下颌齿冠非对称,断面呈椭圆形,舌面沟槽和棱嵴发育,齿冠基部舌侧面发育泡状突。与其

辽宁巨龙

---

* 周长付. 2015. 辽宁巨龙简介(未刊)

他蜥脚类恐龙相似,辽宁巨龙也有长长的脖子、尾部,以及短肥的身体。该化石是本馆2006年发掘的,目前仍在研究中。该恐龙化石的发现不仅进一步丰富了对热河生物群恐龙类群分类多样性的认识,也为了解巨龙形类恐龙的早期演化和辐射提供了新的化石证据。

产地与层位:辽宁北票;早白垩世义县组(距今约1.25亿年)。

2.3.3 鹦鹉嘴龙(*Psittacosaurus*)(幼年群体,"幼儿园"标本)

鹦鹉嘴龙是东亚特有的植食性恐龙,体长一般不超过2米,以群居和育幼行为最具特色。鹦鹉嘴龙头骨吻端高,侧视呈长方形,颧骨突侧向发育,头骨顶视呈三角形,前肢较短,后肢粗壮。本馆展示的是由39个幼体组成的一窝鹦鹉嘴龙化石群体,是迄今已知全世界恐龙幼仔数量最多的一个鹦鹉嘴龙化石群体,被称为"恐龙幼儿园"。这些幼仔大小相当,体长25—40厘米,保存方向基本一致地拥挤在一个0.8平方米的空间里,对研究它们的埋藏情况或生活习性具有重要价值。此外,本馆还展示了一件被称为"母子情深"的标本,示一成年个体的鹦鹉嘴龙和一个幼仔埋藏在一起的"动情时刻"。

产地与层位:辽宁北票;早白垩世义县组(距今约1.25亿年)。

39个幼体鹦鹉嘴龙化石

鹦鹉嘴龙化石及生态复原图

2.3.4 巨齿兽（*Megaconus* Zhou et al.,2013）——迄今最早的带毛发的似哺乳动物

巨齿兽是迄今世界最早的原始形似哺乳动物，模式种为哺乳形巨齿兽（*Megaconus mammaliaformis*），体长约30厘米，体重估计约250克；臼齿有多列的瘤齿，显现了杂食性和植食性。它下颌前臼齿发育一个大而弯曲的齿尖，表明有戳刺能力，可能是为防御而具有的特征。它的下颌式中耳以及原始的踝关节表明了哺乳动物祖先型特征。但巨齿兽的臼齿已高度特化，具有愈合的高冠型齿根和上下牙齿颌精确咬合，表明原始形似哺乳动物已有十分进步的齿形分异和食性的功能适应。巨齿兽的胫骨和腓骨的两端已经愈合，其骨骼特征类似于现生犰狳类或蹄兔类，特别是与非洲现生的岩蹄兔（rock hyrax）的地栖生活方式和习性近似。原始形似哺乳动物很少保存有头颅和肢体骨骼，更少有毛发保存成为化石。巨齿兽是最原始的小贼兽支系（Haramiyida）仅有保存完整的头颅和骨骼，是这一类群化石的十分珍贵的重要发现。

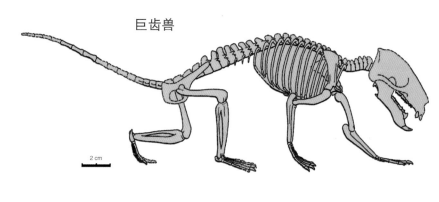

本成果由周长付教授领衔完成，美国罗哲西教授和德国马丁（T. Martin）教授等参与合作。成果于2013年发表于英国《自然》杂志。

产地与层位：内蒙古宁城道虎沟（辽西凌源近邻）；中侏罗世道虎沟组（距今约1.65亿年）。

2.3.5 翔龙（*Xianglong* Li et al.,2007）——世界上最早能滑翔的蜥蜴

翔龙是辽西热河生物群中新发现的蜥蜴类重要分类群，其模式种为赵氏翔龙。在蜥蜴两亿多年的演化历史中，翔龙是唯一发现的、能滑翔的蜥蜴化石。它与生活在东南亚和我国南方的现生飞蜥非常相似，体型小，体长约15厘米，身体两侧有8根加长的肋骨以支撑皮肤翼膜；它既能在树林中攀爬，也能在空中滑翔。赵氏翔龙的发现，填补了滑翔行为在蜥蜴演化史上的空白，也对热河生物群的热带或亚热带气候环境有了更好的提示。

化石由李丕鹏教授带领课题组研究，2007年在《美国科学院院刊》发表。

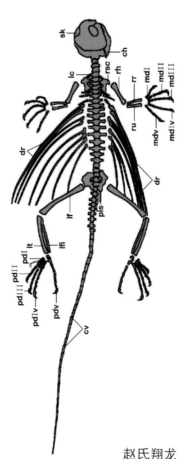

赵氏翔龙

产地与层位：辽宁北票；早白垩世义县组（距今约1.25亿年）。

2.3.6 古果（*Archaefructus* Sun et al.,1998）——迄今世界"最早的花"

古果属是水生草本被子植物，也是迄今已知世界最早的被子植物，包括辽宁古果和中华古果。古果的生殖枝螺旋状着生数十枚蓇葖果，由心皮对折闭合而成，其内包藏着数粒种子（胚珠），柱头未完全分化；雄蕊大多成对状着生，具单沟状花粉，上述特征显示了它们在早期被子植物中的原始性。古果属茎枝细弱，叶子细而深裂，根不发育，未见花瓣和花萼，反映水生性质。据此，孙革等提出被子植物有水生起源的可能性，并首次提出"被子植物起源的东亚中心"假说。上述成果曾先后两次在美国《科学》杂志以封面文章发表，入选"1998年中国十大科技新闻"及美国"2002年百大科学新闻"，有力地推动了我国及全球被

古果
1. 辽宁古果；
2. 中华古果；
3. 中华古果复原图；
4,5. 成果在《科学》上以封面文章发表。

子植物起源与早期演化研究。辽宁古果和中华古果是孙革教授率领课题组1998—2002年于辽西地区首次发现。

产地与层位：辽宁北票、凌源；早白垩世义县组（距今约1.25亿年）。

2.3.7 孔子鸟（*Confuciusornis* Hou et al.,1995）——我国辽西最早发现的原始鸟类

圣贤孔子鸟（*Confuciusornis sanctus*）是我国古鸟类学家侯连海教授等命名的，是辽西最早发现的原始鸟类，属于体型较大的古鸟类。它一方面像今天的鸟类一样，没有牙齿，长有角质喙，尾巴很短并发育尾综骨等；另一方面，它类似于小型兽脚类恐龙，头骨呈双孔型，翅膀上长有锋利的爪，掌指骨未愈合。圣贤孔子鸟的发现填补了自始祖鸟发现以后有

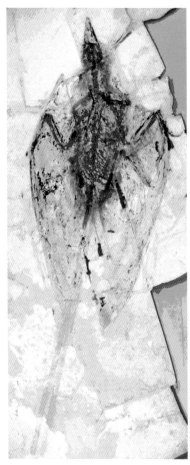

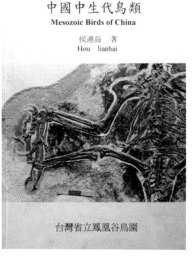

孔子鸟

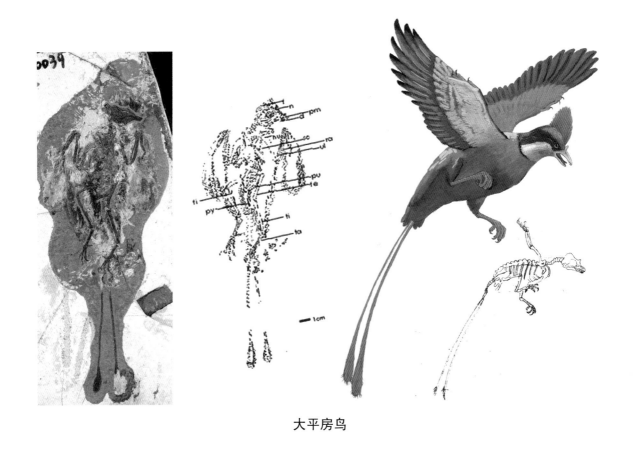

大平房鸟

关早期鸟类研究的重要演化环节的空白。

产地与层位：辽宁北票；早白垩世义县组（距今约1.25亿年）。

2.3.8 大平房鸟(*Dapingfangornis* Li et al., 2005)

大平房鸟是一类中小型古鸟类化石，模式种为棘鼻大平房鸟(*D. sentisorhinus*)。化石由沈阳师范大学古生物研究所2005年于辽宁朝阳县大平房村发掘，李莉副教授率课题组研究并命名。鸟体长(包括尾羽)21.6厘米，头长2.5厘米，羽冠高，鼻骨中部有一突出的棘突，吻尖，牙齿间距较大，额骨低，齿骨长；颈椎异凹型；胸骨发育，具长的后外侧突和短的后内侧突，龙骨突较发育，具小翼羽和两枚长的尾羽，棘突之下方有一不小的孔。棘鼻大平房鸟属于反鸟类始反鸟目。它的发现对研究热河生物群鸟类的分类及演化具有重要意义。

产地与层位：辽宁朝阳大平房；早白垩世九佛堂组(距今约1.2亿年)。

2.3.9 辽龟(*Liaochelys* Zhou, 2010)

辽龟是辽西热河生物群中发现的第3个龟类属，与共生的满洲龟和鄂尔多斯龟有着很近的亲缘关系，但比后二者进步。它们都属于隐颈龟亚目中的原始类群——中国龟科。中国龟科在热河生物群中占有主导地位,以背甲低平、腹甲与背甲以韧带相联等为特征。辽龟化石丰富,主要发现于辽西建昌喇嘛洞地区。其基本特征包括:背甲近椭圆形,长大于宽;椎盾短宽;两个上臀板,第一上臀板明显小于第二上臀板;第三肋板远端扩展。

辽龟由周长付教授研究与命名。

产地与层位:辽宁建昌;早白垩世九佛堂组(距今约1.2亿年)。

辽龟化石

2.3.10 沈师鸟(*Shenshiornis* Hu et al.,2007)

沈师鸟属于鸟类早期演化中的一个地域性特殊类群——会鸟类。会鸟类迄今仅发现于辽西热河生物群中,不仅体型大,而且具有一对超长的前肢,是迄今世界上发现的白垩纪早期个体最大的古鸟类。

原始沈师鸟全长约40厘米,保持"昂首阔步"的体态,它的头骨前部高而粗壮,表现出较德国始祖鸟更为原始的非流线形;其上颌仍保留着原始粗壮的圆锥状牙齿;而后肢的

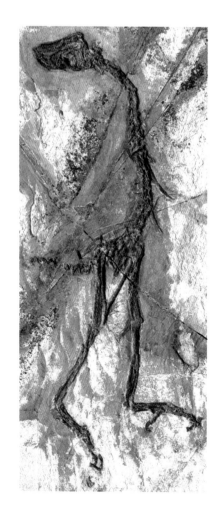

沈师鸟

胫跗骨短,踇趾向后翻转与其他三趾呈对握式,显示出适应树栖生活的特征。原始沈师鸟保存完好的头骨为解决"鸟类可动性头骨早期演化"这一长期争议的难题提供了新的化石证据。研究者们通常认为:在鸟类进化过程中,鸟类可动性头骨的出现较为滞后;与此同时,由于运动系统的改变,行走能力在具尾综骨鸟类的原始类群中已显著削弱。此化石的发现将鸟类可动性头骨和树栖能力演化研究向前大大推进了一步。

原始沈师鸟化石是沈阳师范大学古生物研究所2005年10月在朝阳野外发掘中发现,由胡东宇教授领衔研究。将这一古鸟类新属种命名为"沈师鸟",寓意沈阳师范大学的事业展翅高飞。这一成果2007年发表于《地质学报(英文版)》(*Acta Geologica Sinica*)。

产地与层位:辽宁朝阳大平房;早白垩世九佛堂组(距今约1.2亿年)。

赫氏近鸟龙新标本研究前后

赫氏近鸟龙是一种小型的近鸟类恐龙,2009年初由徐星等命名,模式标本保存在中科院古脊椎动物与古人类研究所,属名"近鸟龙"意为近似于鸟类的恐龙,种名献给了达尔文时代的英国著名生物学家赫胥黎(T. H. Huxley),他曾首次提出"鸟类恐龙起源假说"。近鸟龙的第二件标本(PMOL-B00169)保存在辽宁古生物博物馆,该标本的研究首次证实了带羽毛近鸟恐龙早于迄今最早的鸟类始祖鸟而存在,成果发表在2009年10月出版的英国《自然》杂志上。第三件标本(BMNHC PH828)保存在北京自然博物馆,根据该标本复原了近鸟龙的羽毛颜色,成果发表在2010年3月出版的美国《科学》杂志上。这里讲述的是有关赫氏近鸟龙第二件标本研究的故事。

2008年10月的一天,保存在两片对开石板上的化石摆在了沈阳师范大学古生物研究所标本室的案台上,这就是后来被辽宁古生物博物馆收藏的PMOL-B00169标本。站在标本前,古鸟类研究专家胡东宇的第一印象是"始祖鸟?",因为它与被誉为"最早的鸟"——德国始祖鸟实在是太像了。怀着兴奋和疑问,胡东宇携带标本来到北京、向中科院古脊椎动物与古人类研究所的两位著名专家侯连海(古鸟类学家)和徐星(恐龙学家)请教,这两位科学家分别是沈阳师范大学特聘和双聘教授。

通过比较,侯、徐、胡等三位专家都感到沈阳师范大学的新标本与已知辽西产的带羽毛恐龙和鸟类均有很大的差异。那么这个带羽毛物种到底是"龙"还是"鸟"? 开始,侯先生倾向于"鸟",因标本具有很多与始祖鸟相似的特征。徐星则倾向于"龙",因标本也表现出与伤齿龙类相似的特征。实际上,这种争议可能涉及"鸟"的概念问题:如果将始祖鸟定为"鸟",近鸟龙也应是鸟;但如果将始祖鸟定为"龙",则近鸟龙也应是龙。

由于利用现代系统分析手段可以避免争议,三位专家开始对新标本做系统分析,而结果显示:赫氏近鸟龙处在原始伤齿龙类位置。也就是说,新标本是一个比始祖鸟还要早的、处于飞行演化初期的带羽毛恐龙,即是连接恐龙向鸟类演化的"桥梁"。于是,根据系统分析和同世界范围内发现的

胡东宇(右)与侯连海在沈阳师范大学

徐星在介绍赫氏近鸟龙化石

2.4 闪亮的科研

辽宁古生物博物馆一直将科学研究作为检验博物馆总体水平的重要标志。根据博物馆科研人员的自身优势和地域优势,建馆五年来,全馆突出将古植物学(重点为早期被子植物)、古鸟类、古两栖爬行类、古哺乳类以及辽宁古人类等5个研究领域作为主攻方向,瞄准国际前沿,产出了一批重要成果,发挥了科研在博物馆运行中的领军作用。

2.4.1 主要研究进展

几年来,辽宁古生物博物馆的科研人员共发表论文100余篇,含SCI检索论文40余

确凿化石的比较,徐星和胡东宇等修订建立了一个新的兽脚类恐龙系统发生树,提出了兽脚类恐龙的所有主要类群在晚侏罗世最早期之前可能都已出现并迅速分化,包括鸟类在内的许多重要类群处于这次迅速演化事件中出现的"兽脚类恐龙分异的时间框架"中。与此同时,有关化石所在地层及其时代研究也传来了喜讯:新标本产地——辽宁建昌玲珑塔大西山的野外工作表明,这里的地层属髫髻山组,同位素测年值在1.61亿年左右,即赫氏近鸟龙生活时代应在中侏罗世晚期至晚侏罗世早期,比德国始祖鸟要早至少1000万年!这意味着在中国辽西找到了一个比"迄今最早的鸟"还要早的带羽毛物种。而室内的标本形态对比研究也支持野外调查结果。对标本的观察发现,在飞行能力演化上,近鸟龙比始祖鸟原始:虽然近鸟龙的前肢伸长、相对长度与始祖鸟的相似,但始祖鸟的前肢下臂用来附着飞羽的尺骨已像鸟类一样向后拱曲,更有利于翼扇面的形成,而近鸟龙的尺骨仍是直的;始祖鸟翅膀上的飞羽已像鸟类一样,羽轴前侧羽片窄,后侧羽片宽,更适于飞行,而近鸟龙的飞羽相对窄,且羽轴纤细,羽轴两侧的羽片仍是对称的。多方面综合研究完全肯定了最初的判断,徐星和胡东宇等有了足够信心证实:带羽毛近鸟恐龙早于始祖鸟而存在。

近鸟龙的第二件化石标本 PMOL-B00169

历时6个月室内研究终于有了可喜收获。2009年10月1日,正当国庆60周年大庆时,英国《自然》杂志正式发表了胡东宇等有关赫氏近鸟龙研究的这一新成果。看着天安门广场威武雄壮的阅兵式直播,胡东宇和徐星等自豪地在想:在古生物界他们又有一份珍贵的礼物献给祖国。后来,这一成果先后入选了2009年"中国高校十大科技进展"、"中国十大科技进展新闻"和美国《科学新闻》杂志评选的"2009年世界科学新闻"等。2010年赫氏近鸟龙标本还代表辽宁出展于"上海世博会"。英国《自然》杂志还请徐星9月25日在英国举办的国际古脊椎动物学术年会上提前做了此项新发现的报告。中国辽宁侏罗纪"最早的带羽毛恐龙的发现"的消息迅速传遍世界。

[节选自胡东宇《大自然》(2014)文章,部分修改]

寻找"第一朵花"的故事

被子植物（也称有花植物）是现今植物界最高级的类群，现有约400个科30万种，几乎遍布全球。它们具有真正的花，种子有果实包藏，先进的繁殖方式使它们在植物界最为繁盛。但世界上最早的被子植物（"花"）出现在哪里？它们又是如何起源和演化的？这一系列问题一直困扰着科学家们。100多年前，著名生物学家达尔文称之为"讨厌之谜"。

1996年秋，时任中科院南京地质古生物研究所研究员的孙革收到刚从辽西回来的同事送来的3块标本，化石产自辽西北票黄半吉沟。当孙革看到第三块化石时，被震惊了：该标本是一个很像蕨类的分叉状枝条，但似叶片的部分呈凸起状，显然是种子；主枝和侧枝上螺旋状排列着40多枚豆荚状的果实（蓇葖果），每个蓇葖果中包藏着2~4粒种子。这些是典型的被子植物特征！由于化石产于辽西著名的义县组，显然是全世界被子植物化石产出的最低层位。孙革马上意识到，这是迄今全世界最早的被子植物，并建议起名"辽宁古果"。当晚，他把这一消息电话通知了老朋友、课题组

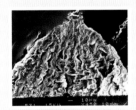

辽宁古果

的同事——古植物学家郑少林，两人都十分兴奋。找寻"第一朵花"的研究工作便从此开始。

古生物学家的工作原则是，研究化石标本必须准确地掌握它们的产地和层位。为做好这项工作，课题组郑少林和他夫人张武亲自赶到北票黄半吉沟，晚上住在老乡家，白天上山找化石。功夫不负有心人，张武很快先发现了一块"辽宁古果"的果实碎片，紧接着郑少林又发现两块较完整的蓇葖果化石。三件化石立即被送往南京，孙革喜出望外。由于新发现的蓇葖果化石保存了角质层，孙革在实验室里亲自做分析实验，发现了它们的种皮的表皮细胞：略弯曲的垂周壁有些角质加厚，看上去似反映曾受到干旱等古环境影响。春天到了，孙革率课题组一起赶到北票黄半吉沟采化石、测剖面，一共采得17件"辽宁古果"标本。夏天，郑少林又去采集，又发现两块更重要的标本：原来他以为是松柏类，但孙革结合经验敏锐地认识到：这正是苦苦要寻找的"辽宁古果"的雄蕊化石！孙革将自己关在实验室几天，终于从对开保存的26枚花药中发现了原位花粉：花粉很小，长轴大约30微米，具单沟，是典型的被子植物花粉！有了雄蕊，又有了雌蕊，"辽宁古果"这一早期被子植物的特征越发清楚显现。正在这时，课题组的周浙昆研究员陪植物学界的泰斗吴征镒院士来南京，吴老亲自在实验室看了"辽宁古果"标本，确认它是被子植物，而且认为肯定是一个新类群！

在周浙昆协助下，孙革很快完成了论文"迄今最早的被子植物在中国的发现"；为了向国际前沿冲刺，孙革决定将文稿投到国际权威学术刊物——美国《科学》杂志。但为保证和提高论文的学术质量，孙革决定还是将文稿先寄给老朋友、国际著名古植物学家、美国科学院院士迪尔切，请他在分类方面予以指导。没几天，迪尔切院士热情回复孙革：他认为辽宁古果在总体分类上似更接近于木兰科，建议将分类部分进一步改写和补充，并支持向美国《科学》投稿，因为这是全世界古植物学领域的一项新的重大发现。

此时正是1998年的春天，温暖和喜悦交融在南京、沈阳、昆明和美国。夏天里，迪尔切院士和孙革等又一起赴北票黄半吉沟野外，并又有一些有关"辽宁古果"的新发现。11月27日大洋彼岸传来喜讯：美国《科学》杂志以封面文章发表了孙革等撰写的《追索最早的花——古果》的论文！一时间，美、英、日及新加

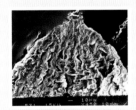
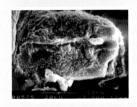

辽宁古果种皮角质层和花粉　　26枚花药　　孙革（左）与郑少林在南京（1998）

篇。其中在国际权威学术刊物英国《自然》杂志发表论文6篇(含第一作者3篇),《美国科学院院刊》2篇,多篇论文在美国《古脊椎动物学杂志》(*JVP*)、美国《国际植物学杂志》(*IJPS*)、英国《BMC进化生物学》(*BMC Evolutionary Biology*)、《中国科学(英文版)》(*Science China*)等重要刊物上发表。出版了《30亿年来的辽宁古生物》(2011)、《黑龙江嘉荫晚白垩世—古新世生物群、K-Pg界线及恐龙灭绝》(2014)、《华北板块东部早古生代动物群、沉积相及地层多重划分》(2015)及《中国蕨类植物画》(2015)等专著4部,教材2部。五年来获国家自然科学基金项目及省部级专项基金项目等36项,科研总经费800余万元;荣获辽宁省科学技术奖(自然科学)一等奖、二等奖及三等奖各1项;"迄今最早的带羽毛恐龙——赫氏近鸟龙的首次发现"成果入选教育部"中国高校十大科技进展"、两院院士评选的"2009世界/中国十大科技进展新闻",为我国及全球鸟类起源及早期演化研究作出了突出贡献;赫氏近鸟龙标本应邀在上海世博会展出,受到国土资源部的嘉奖。

主要研究成果包括首次发现迄今世界最早的带羽毛恐龙——赫氏近鸟龙(胡东宇等,2009),以及侏罗纪原始形似哺乳动物——巨齿兽的发现(周长付等,2013),沈师鸟(胡东宇等,2010)、渤海鸟(胡东宇等,2012)、盛京鸟(李莉等,2012)、翔鸟(胡东宇等,2013)等一批古鸟类的发现;黑龙江嘉荫K-Pg界线确定(孙革等,2011,2014),我国侏罗纪最大恐龙——新疆巨龙的发现(吴文昊、周长付等,2013),辽西侏罗纪真蕨类紫萁科矿化根茎化石新发现(田宁等,2013,2014),龟类及翼龙类研究进展(周长付,2012,2015)等。

坡等20多个国家近百家新闻媒体纷纷以"世界最古老的花在中国"为主题报道了来自中国的重大发现;美国NOVA公司还专门制作了专题片《第一朵花》(*First Flower*)向全球播放。

辽宁古果以其特有的被子植物特征,展示了距今1.25亿年原始的被子植物的面貌,促使孙革等提出"被子植物起源的东亚中心"新假说,并提出了被子植物可能水生起源的新的研究思路。国际著名植物学家、前美国植物学会主席瑞温(P. Reven)院士认为"古果属(*Archaefructus*)是迄今已知世界上最古老的花"(2005);我国植物学家吴征镒院士认为"辽宁古果的发现为中国人解决植物学界的重大理论问题作了重要贡献"(2003)。2012年国际著名被子植物专家道伊尔(J. A. Doyle)将古果属和义县组作为全球早期被子植物演化的底界。

辽宁古果,这个古老的"花的使者",已将它的芬芳撒向世界。

《科学》1998年封面　　迪尔切院士(右)与周浙昆教授　　"第一朵花"专题片(2006)

近五年发表的部分论文的期刊封面

所获奖项及部分新出版的专著

2.4.2 部分新发现补充记述

2.4.2.1 反鸟类新类群（翔鸟、盛京鸟）的发现

反鸟类新属种神秘翔鸟（*Xiangornis shenmi*）的发现使人们对反鸟类以及整个早期进步鸟类的飞行演化有了一个全新的认识。翔鸟的发现表明，腕掌的飞行形态特征在反鸟类和今鸟类中可能是独立演化的，并且在反鸟类中可能更早地出现。该成果由胡东宇等于2013年在国际古脊椎动物研究权威期刊——美国《古脊椎动物杂志》上发表。

翔鸟及发表杂志

杨氏盛京鸟（*Shengjingornis yangi*）发现于锦州头道营子早白垩世九佛堂组（距今约1.2亿年），属于反鸟类中形体较大的类群，具有长而前端弯曲的喙，上下颌仅前端具少数牙齿，前后肢近等长，它们可能和现生鸟类中的鸻鹬鸟类一样，以探挖滩涂泥沙中的蚯蚓、沙蚕等为食，属于营涉水生活的鸟类。新发现为揭示反鸟类的一支曾向近水生活演化提供了化石证据，并为长翼鸟类研究提供了新的解剖学信息。成果由李莉等2012年发表于《地质学报（英文版）》。

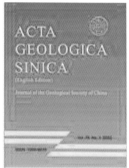

盛京鸟及发表杂志

2.4.2.2 建昌颌翼龙头骨新发现

翼龙类是爬行动物中最早成功飞上天空的一个类群。建昌颌翼龙头骨的发现揭示了侏罗纪建昌颌翼龙重要的头骨形态特征，不仅丰富完善了已知建昌颌翼龙骨骼形态特征，而且揭示了其头骨形态特征的生长发育变化。论文还通过与德国索洛霍芬古动物群对比，探讨了我国东北非翼手龙类的分异度和生态习性。该成果由周长付2014年在美国《古脊椎动物杂志》上发表。

建昌颌翼龙头骨化石及发表杂志

2.4.2.3 黑龙江嘉荫K-Pg界线确定

首次在黑龙江嘉荫以"厘米级"精度确定了我国首个具有国际对比标准的陆相K-Pg界线;并提出火山活动、海平面下降和气候变冷等可能是嘉荫乃至整个东北亚地区恐龙灭绝的主要原因,以及黑龙江流域的恐龙灭绝可能早于K-Pg界线时间的新观点。成果对研究我国及东北亚地区中生代与新生代之交生物群、地质事件、地层划分对比、特别是了解恐龙灭绝等重大科学问题,具有重要意义。该成果由孙革等2011年在《世界地质》(*Global Geology*)及2014年出版的专著中发表。

2.4.2.4 侏罗纪真蕨类紫萁科矿化根茎化石新发现

辽西侏罗纪真蕨类紫萁科矿化根茎化石新发现,分别建立两个新种——北票阿氏茎(*Ashicaulis beipiaoensis*)和王氏阿氏茎(*A. wangi*),为探讨紫萁科植物在北半球的宏演化提

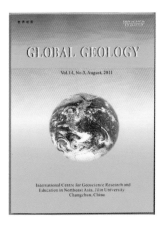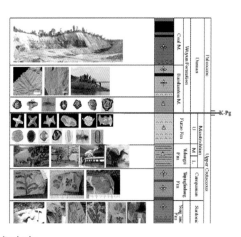

黑龙江嘉荫K-Pg界线确定

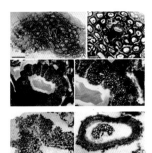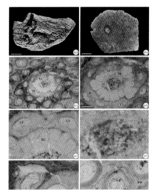

侏罗纪真蕨类紫萁科矿化根茎化石

供了新证据。该成果是由田宁等于2013年在美国《国际植物学杂志》和2014年在《中国科学：地球科学(英文版)》(*Science China Earth Sciences*)上发表。

2.5 活跃的科普

2.5.1 丰富多彩的科普活动

"科学面向大众、服务大众"是辽宁古生物博物馆始终坚持的理念。五年来，博物馆大力开展科学普及活动，特别是加强有关"生命进化"知识的传播，迄今已累计接待观众近百万人次，节假日最多曾达到7000人/日，接待各界社会团体近4000个；先后建立了71所中小学科普合作校，招募了五批共800余名大学生志愿者；开展了"小小讲解员"培训，以及"博物馆奇妙夜"、"青葱小课堂"、"走进社区科普大学"、"博物馆少儿科普剧表演大赛"等活动；结合、博物馆日、科普日、世界地球日、环境日等重大纪念日，开展了"我与恐龙是朋友"、"认识自然、采集标本"、"爱护环境、珍惜地球资源"等主题鲜明的纪念活动。

为加强科普宣传，博物馆出版了《30亿年来的辽宁古生物》(孙革等，2011)、《走进鸟的故乡》(孙革等，2013)、《探秘辽宁史前世界》(杨建杰等，2015)及《龙鸟传奇》(杨建杰等，2015)等科普著作。孙革馆长等撰写的《30亿年来的辽宁古生物》被评为2014年全国国土资源系统优秀科普图书，获2014年辽宁省国土资源系统优秀科技成果一等奖等；杨建杰副馆长主编的《生命从远古走来》荣获2016年全国国土资源系统优秀科普图书。

看特展

小小讲解员

化石讲解

少儿科普剧表演大赛

走进社区科普大学堂

"科普合作校"授牌

被授予的部分科普基地称号

五年来，辽宁古生物博物馆先后被授予"中国科协全国科普教育基地"、"中国古生物学会全国科普教育基地"、国土资源部"国土资源科普基地"、"辽宁省科学技术普及基地"、"辽宁省环境教育基地"、"沈阳市科普基地"、"沈阳市爱国主义教育基地"、"沈阳市环境教育基地"及"沈阳市青少年科技教育基地"等10个省市级以上的科普基地称号；被评为沈阳市教科工委系统"服务沈阳工作先进集体"（2012）、由中国自然博物馆协会评选的"全国自然类博物馆优秀集体"（2015），并荣获2014年沈阳市文物局评选的沈阳市首届博物馆陈列展览"精品奖"等。活跃的科普活动为提高辽宁广大群众的科学文化素养、推动全国古生物科普工作作出了积极的贡献。

2.5.2 举办特展

举办特展，是辽宁古生物博物馆在科普活动中的亮点之一。为更好地开展科普工作，充分利用辽宁的化石资源优势和博物馆的人才优势，自2011年以来，博物馆先后举办了"辽宁恐龙特展"（2012）、"辽宁古植物化石特展"（2013）、"马化石特展"（2014）及"从猿到人特展"（2016）等4次较大型的特展活动，并在法国、德国等自然史博物馆多次举办了辽宁化石特展（参见2.6.3）。

"辽宁恐龙特展"主要介绍了辽宁带羽毛恐龙的发现及其在研究全球鸟类起源中的意义。我国著名地质学家、中国科学院院士李廷栋，著名古生物学家侯连海、徐星、姬书安等，国家古生物化石专家委员会办公室副主任王丽霞，中国古生物学会秘书长王永栋，辽宁省国土资源厅前副厅长杨德军、省化石局管理局前局长孙永山、吉林大学古生物研究中心主任孙春林以及沈阳师范大学副校长李铁君等应邀出席；辽宁省与沈阳市各界来宾及

发表的科普图书、宣传品

辽宁恐龙特展（2012）
1. 特展开幕仪式主席台；
2—5. 李廷栋院士（2）、王丽霞副主任（3）、哥德弗洛依特教授（4）、李铁君副校长（5）等来宾致辞；
6. 特展展厅；
7. 徐星教授亲自讲解。

百余名观众出席了开展仪式。

"辽宁古植物化石特展"是为庆祝第37个"国际博物馆日"而举办,并作为"2013年辽宁省暨沈阳市科技活动周"主体活动之一。沈阳市政府副秘书长徐兴家等领导亲临特展现场,3000余名观众兴致勃勃地参加了本次活动。本次特展是我国首次举行的有关古植物化石的专题展览,重点介绍植物起源与进化以及古植被在辽宁的发展进程。特展以丰富而精美的化石实证,回溯了4亿多年以来地球植物界的发展与演化历程,呼吁人们热爱大自然、保护生态环境、促进辽宁古生物化石保护工作。

2014年时值中国传统的马年,4月20日,为纪念第45个"世界地球日",由辽宁省国土资源厅主办、辽宁古生物博物馆承办了"珍惜地球资源,2014马化石特展",主要宣传马的进化及马与人类的密切联系,强调人与大自然的和谐发展,珍惜地球资源与环境。此次展览的展品包括了德国麦索的小原始马(复制品)、我国发现的贾氏三趾马、平齿三趾马、埃氏马、普氏野马等完整骨架化石,以及大连马、中华马、云南马、三门马等共9件化石珍品,介绍了5800万年以来马的进化史、辽宁及我国马化石分布、马与人类和谐相处的历史以

辽宁古植物化石特展(2013)

1,2. 特展展厅及观众参观;3. 沈阳市政府副秘书长徐兴家(左5)与林群校长(右5)出席特展开幕式;4. 徐兴家副秘书长(中)等参观特展。

马化石特展(2014)
1. 省国土资源厅副厅长杨旭致开幕词；2. 傅仁义教授(右3)向小观众介绍马化石；3. 王丽霞副主任(右1)等在展厅；
4. 英国爱德华院士(左2)参加特展活动。

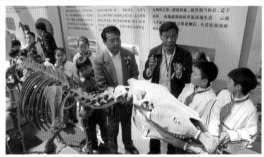

及"马文化"等，是我国首次会聚中外马化石展示为一体的"化石盛宴"。国家古生物化石专家委员会顾问、中国科学院李廷栋院士，英国皇家学会会员、伦敦林奈学会主席爱德华(D. Edward)院士，国家古生物化石专家委员会办公室副主任王丽霞，中国古生物学会秘书长王永栋，辽宁省国土资源厅副厅长杨旭、马原，沈阳师范大学副校长郝德永，著名古脊椎动物学家徐星，以及部分国内自然类博物馆与化石管理部门负责人及学生代表等200余人出席了开幕仪式。此次特展得到中国古生物学会、辽宁省地质学会、中国科学院古脊椎动物与古人类研究所、吉林大学古生物研究中心、大连自然博物馆、锦州古生物博物馆等的大力协助和支持；《国土资源报》等近百家新闻媒体相继报道了本次特展的盛况。

2016年5月，为庆祝第40个"国际博物馆日"，结合我国传统"猴"年，博物馆与辽宁省国土资源厅及沈阳市科技局等联合主办了"从猿到人特展"。特展开幕式上，我国著名地质学家李廷栋和刘嘉麒院士，国家古生物化石专家委员会办公室副主任王丽霞，国土资源部重点实验室办公室副主任张辉旭，中国古生物学会秘书长王永栋，辽宁省国土资源厅副

从猿到人特展(2016)
1. 特展揭幕(左起:张辉旭、王大超、李廷栋、刘嘉麒、马原、王丽霞);
2. 日本水田宗子理事长(右1)在沈阳师范大学于文明书记(右2)陪同下参观特展。

厅长马原、沈阳师范大学副校长王大超,以及国内专家学者及师生等百余人出席。当天下午,日本著名教育家、城西大学理事长水田宗子(Noriko Mizuta)教授在沈阳师范大学党委书记于文明陪同下,也兴致勃勃地参观了展览,对该特展的举办予以高度称赞。

此次特展介绍从古猿到人的演化发展历程,特别介绍了辽宁的古人类起源、演化及化石分布,这是我国东北地区首次会聚地史时期中外灵长类化石的一次"盛宴"。特展展品主要包括德国麦索的达尔文猴、意大利的山猿、非洲埃塞俄比亚的南方古猿露西(被誉为"人类祖母")、最大的灵长类动物——步氏巨猿,著名北京猿人头盖骨、东北地区最完整的直立人——金牛山人,以及一批伴随古人类生存的哺乳动物化石等30余件化石珍品及复制品。

2.5.3 专家参与指导

众多专家亲自参与及指导、专家走近观众,是辽宁古生物博物馆特展活动的特色之一,也使特展的科普活动具有更深厚的科学底蕴。2012年5月举办的"辽宁恐龙特展"特邀了我国年轻著名恐龙学家、近鸟龙及小盗龙命名人徐星,古鸟类学家、孔子鸟命名人侯连海,以及古脊椎动物学家、中华龙鸟命名人(之一)姬书安等专家,他们不远千里来到沈阳,亲自向观众介绍辽宁带羽毛恐龙的发现对研究鸟类起源的科学意义等,给参加特展的观众,包括馆内工作人员带来极大的惊喜!2014年举办"马化石特展"时,特邀了我国著名古哺乳动物学家邓涛教授从北京赶来,提前到会指导,并做了关于"马的起源与进化"的专题报告,将特展的学术水平推向国内一流。古生物学家董枝明、段吉业、郑少林、张立君等老一辈专家多次亲临特展现场指导。2012年,国际著名恐龙学家、比利时皇家自然史博物馆的哥德佛罗伊特教授应邀出席了"辽宁恐龙特展"。2014年"马化石特展"举办时，国际著名古植物学家、英国林奈学会主席爱德华院士也在百忙中从北京赶来出席，还参加学生们组织的科普节目，使特展活动充满着国际化的氛围。国内外"大家"们亲临现场，使参会群众真正尝到了一份份精美的"科普大餐"。

专家亲临特展现场指导（2014）
1. 徐星教授讲带羽毛恐龙；2. 侯连海教授在特展现场；3. 邓涛教授（左2）指导马展化石布局；
4. 刘嘉麒院士（左2）参观马化石特展；5. 英国爱德华院士（左3）和李廷栋院士（中）出席特展；
6. 孙革馆长（左2）亲自作解说。

2.5.4 科普会议与培训班

为推动我国地质古生物科普工作开展，受中国古生物学会科普工作委员会的委托，辽宁古生物博物馆于2011年9月承办了"首届全国地质古生物科普工作研讨会"，来自包括台湾的全国30多家自然类博物馆及国家地质公园等单位的代表等近百人参加，国家古生物化石专家委员会办公室副主任王丽霞、中国古生物学会秘书长王永栋、中国科学院南京地质古生物研究所副所长王海峰，以及著名恐龙学家徐星等专家应邀出席。研讨会有力地加强了与会各单位间的交流学习，推动了我国地质古生物科普工作的蓬勃开展。

为促进我国地质古生物博物馆领导工作水平的提高，也包括馆长们在领导层面上专业水平的提高，在国家古生物化石专家委员会办公室的大力支持下，博物馆与中国古生物学会于2013年和2015年在沈阳联合举办了两届"全国地质古生物博物馆馆长专业培训班"，取得显著效果。

首届全国地质古生物科普工作研讨会（2011.9）
1. 冯伟民常务副主任主持会议；
2. 沈阳师范大学王大超副校长致辞。

2013年1月在沈阳举办的"首届全国地质古生物博物馆馆长专业培训班"上，包括中国科学院刘嘉麒、舒德干、周忠和等院士以及中国古生物学会理事长杨群在内的14位著名地质古生物学家及相关领域专家，就地质古生物学基础知识、国家化石保护的法律法规

首届全国地质古生物科普工作研讨会全体合影（2011.9）

首届全国地质古生物博物馆馆长专业培训班（2013.1）
1. 全体合影；
2. 沈阳师范大学林群校长在开班仪式上致辞；
3. 辽宁省化石保护局前局长孙永山致辞；
4—6. 刘嘉麒（4）、舒德干（5）、周忠和（6）等院士授课；
7. 学员在北票野外实习。

第二届全国地质古生物博物馆馆长专业培训班（2015.1）
1. 全体合影；2. 培训班组委会主任孙革馆长主持开班仪式；3. 中科院殷鸿福院士致辞；
4. 学员听课现场；5. 学员们在化石修复室实习。

以及博物馆管理等，进行了高水平的授课；我国古生物学家孙革、万晓樵、金昌柱、王伟铭、徐星、王永栋，博物馆管理专家孟庆金、冯伟民、王原，以及化石保护管理专家王丽霞等，都在百忙中前来授课。来自全国30个省、市的50个地质古生物博物馆和化石保护管理部门的代表参加。培训班对提高我国地质古生物博物馆和化石管理的总体水平以及更好地开展科普工作发挥了积极作用。

2015年1月，应国内部分自然类博物馆馆长及相关部门的建议，辽宁古生物博物馆在沈阳举办了"第二届全国地质古生物博物馆馆长专业培训班"。我国著名地质古生物学家殷鸿福、刘嘉麒、周忠和院士等共15位专家以及国家古生物化石专家委员会办公室副主任王丽霞等前来授课。来自国内14个省、直辖市的地质古生物博物馆和化石保护管理部门的馆长或代表参加了培训。培训班的开办受到了广泛欢迎；有关领导称赞辽宁古生物博物馆此项在馆长层面上开展的科普工作可以成为"品牌"活动。

热河生物群的龟类化石

龟类一般身裹坚硬的甲壳，头和四肢可以缩入壳内。龟类是一类古老而特殊的爬行动物，化石最早发现于距今约2.2亿年的晚三叠世。依据脖子收缩方式的不同，现生龟类分为侧颈龟类和隐颈龟类两大类群。其中，隐颈龟类最丰富，广泛分布世界各地；鳖也属于此类，只是缺少角质壳和骨化的"裙边"。

热河生物群的龟类大都属于泛隐颈龟类的原始类群，如满洲龟（*Manchurochelys*）、鄂尔多斯龟（*Ordosemys*）、辽龟和小龟（*Xiaochely*）等，都以龟甲低平、背腹甲以韧带相联等为特征。热河生物群还发现有鳖类化石，如连鳖（*Perochelys*）等。

热河生物群的龟类化石展示了许多以往我们了解甚少的生态习性信息。现生陆龟的脚往往短粗，方便陆地爬行；水生龟（如猪鼻龟、海龟等）的脚掌往往加长，呈桨状或鳍状，适于划水。根据测量对比，热河生物群的龟类可能都属于水生龟，且水生习性强；此外，它们往往都有一个长尾巴。现生龟的尾巴大都很短，只有北美的鳄龟、我国和东南亚地区的大头龟（鹰嘴龟）有长尾巴。鳄龟的尾巴可能用于协助水底行走；大头龟的尾巴则用于攀爬中的支撑，据说也可能在打斗中起到恐吓对方的作用。虽然二者在亲缘关系上相距很远，但它们都是比较凶猛的龟，并且头部都比较大，不能缩入壳内。这些也为我们推测热河生物群龟类的长尾巴的功能，以及其他生态习性提供了更多思考。

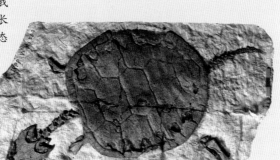
满洲龟

辽龟

2.6 国内外交流

由于古生物研究的国际化特点和博物馆发展的国际化趋势,辽宁古生物博物馆始终坚持国际化的办馆方向。开馆以来,接待了美国、德国、俄罗斯、英国、法国、奥地利、日本、韩国、朝鲜、蒙古、越南、泰国、比利时、罗马尼亚、加拿大、印度、巴西、以色列及爱尔兰等20余国百余人次来访开展学术交流,举办了三次大型国际学术会议。与德国、美国、俄罗斯、比利时及韩国等国开展了多项合作研究项目,合作发表了20余篇论文。与此同时,博物馆还多次赴德、法等国联合举办古生物化石展览,宣传我国、特别是辽宁在古生物研究、化石保护及博物馆建设等方面取得的成绩。为支持辽宁古生物博物馆建设,美国科学院迪尔切院士将毕生珍藏的6000余册专业图书和3万多册单行本无偿捐献给辽宁古生物博物馆,博物馆为此成立了"迪尔切图书室";德国孢粉学家阿什拉夫(A. R. Ashraf)教授捐赠

博物馆成员赴国外学术交流
1. 博物馆10名成员出席在德国格丁根举办的中德古生物会议;
2. 胡东宇赴奥地利出席国际会议;
3. 张宜访问以色列海法大学;
4. 段冶、赵鑫赴俄罗斯贝加尔博物馆出席国际会议。

迪尔切图书室和阿什拉夫实验室

1. 迪尔切图书室；2. 迪尔切院士在馆作报告；3. 阿什拉夫实验室；4. 阿什拉夫教授（左）指导实验室工作。

"东北亚古生物学协同创新中心"签字仪式（2014）

中科院南京地质古生物研究所所长杨群（一排右7）、沈阳师范大学校长林群（一排左7）
及中科院院士李廷栋（一排左6）等在辽宁古生物博物馆出席签字仪式。

了3000余册专业书籍和显微镜等设备,博物馆为此建立了"阿什拉夫实验室"。

在国内,博物馆所在沈阳师范大学先后与中科院南京地质古生物研究所及古脊椎动物与古人类研究所签订了"东北亚古生物学协同创新中心"合作协议,为博物馆的工作起到引领作用。博物馆还与北京自然博物馆,南京古生物博物馆、深圳古生物博物馆、中国古动物馆、自贡恐龙博物馆、大连自然博物馆、重庆自然博物馆、河南地质博物馆、安徽地质博物馆、新疆地质博物馆、黑龙江嘉荫恐龙博物馆、吉林大学地质博物馆及河北地质大学地质博物馆等,开展了经常性的馆际学术交流。所有这些交流与合作使博物馆的工作能借鉴更多兄弟博物馆的宝贵经验并促进协作。

2.6.1 国际项目合作

五年来,博物馆与德、俄、美、比、韩等国在有关中生代、新生代地层古生物的合作项目研究工作取得多项重要成果。2011—2013年开展的《吉林东部中新世长白植物群及与欧洲对比》项目(中德科学基金项目CZ654项目资助)首次提出吉林长白八道沟植物群的时代是上新世,大大提高了我国东北新生代生物地层的研究精度,并促进了博物馆与斯图加

国际合作项目

1. 中德新疆中生代野外联合考察(2013); 2. 中德长白中新世植物群合作(2013);
3. 中俄吉林野外合作考察(2015); 4. 中比恐龙合作(2011)。

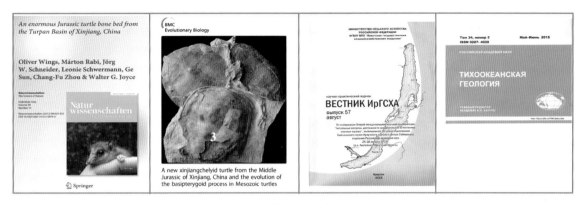

国际合作研究新成果

特自然史博物馆在德国联合举办辽宁化石展的顺利进行。国家自然科学基金资助的中德新疆中生代地层合作项目多年来也取得多项重要成果,波恩大学及图宾根大学等在鄯善新疆巨龙及乌苏新疆龟等重要发现研究中,都作出贡献。由中比科技合作项目支持的"中国东北部白垩纪温室效应"合作项目,比利时皇家自然博物馆恐龙专家哥德弗洛依特教授及其领导的课题组在曙光鸟等研究中取得突出成绩,与博物馆胡东宇教授在《自然》上合作发表2篇论文。此外,中俄在黑龙江嘉荫白垩纪生物群及K-Pg界线研究合作中取得重要进展;中美在早期被子植物合作研究中也取得可喜成绩。

2.6.2 国际学术会议

2011年8月辽宁古生物博物馆/沈阳师范大学古生物学院与伊春市政府及吉林大学在黑龙江伊春联合主办了"伊春地质古生物国际会议",来自美国、俄罗斯、英国、德国、法国、日本、印度、以色列等15个国家百余名专家学者出席。中国地质学理事常务副会长孟宪来、中国古生物学会理事长杨群、国土资源部环境司副司长陈小宁、黑龙江省副省长于莎、著名地质学家中国科学院院士李廷栋、美国科学院院士迪尔切、俄罗斯通讯院士阿克米梯耶夫、德国阿什拉夫教授以及国际地层委员会副主席彭善池等应邀出席。孙革馆长代表国家自然科学基金重大国际合作项目组,在会上正式宣布了在黑龙江嘉荫首次确定我国陆相白垩纪—古近纪地层界线(K-Pg界线)的新的重大成果,得到与会国内外专家的一致肯定和高度评价。8月24日,与会代表在嘉荫小河沿隆重举行K-Pg界线立碑典礼,整个会议活动达到了高潮。

2015年8月16—20日,由辽宁古生物博物馆与中科院南京地质古生物研究所及中国地质大学(北京)、吉林大学古生物研究中心联合主办的"第12届中生代陆地生态系统国际会议"(MTE-12)在沈阳隆重举行。中国科学院李廷栋院士,国际古生物学会主席周忠

伊春地质古生物国际学术研讨会(2011.8)

1. 俄罗斯专家；2. 中外专家及国家古生物专家委员会代表(左起)王丽霞、迪尔切、李廷栋、孙革、李继江；
3. 开幕式；4. 嘉荫小河沿 K-Pg 界线铸点立碑典礼。

和院士，国家古生物化石专家委员会办公室副主任王丽霞，俄罗斯自然科学院院士基里洛娃，美国佛罗里达自然史博物馆馆长邛斯(D. Jones)，英国开放大学教授斯派瑟(R. A. Spicer)，德国波恩大学地质古生物研究所前所长马丁及阿什拉夫教授，韩国古生物学会主席许民(M. Huh)，印度沙尼古植物研究所所长巴依佩(S. Bajpai)，国际地科联 IGCP-608 项目主席安藤寿男(H. Ando)，日本中央大学教授西田治文(H. Nishida)，以及来自18个国家的160余名专家代表出席了会议，中国古生物学会副理事长、馆长孙革教授担任会议主席。会议包括4个大会主题报告和110个口头报告及墙报，对中生代生物多样性、地质环境、鸟类起源与恐龙演化、古植物、古气候及古生物足迹等进行了深入研讨。代表们还赴辽西朝阳、建昌等进行了野外考察。会议举办得到国内外专家一致好评。

除在馆内及国内开展的国际学术交流外，博物馆成员近年来赴德国、日本、俄罗斯、美国、奥地利、以色列等国出访交流及出席国际学术会议20余人次，2014年9月孙革馆长还应邀赴俄罗斯远东国立大学讲学等。

第12届中生代陆地生态系统国际学术研讨会（2011.8）
1. 与会者合影；2. 中方领导及专家（左起）王丽霞、周忠和、马原、李廷栋、孙革、郝德永；
3. 研讨会上；4. 大会会场；5. 辽西建昌野外考察。

2.6.3 国际化石展

为扩大国际交流与合作，将古生物科普活动扩展到国外，在国家古生物化石专家委员会办公室和辽宁省国土资源厅的大力支持下，博物馆与吉林大学古生物研究中心等于2012、2014及2015年赴法国和德国，成功联合举办了三次古生物化石展览，宣传了我国，特别是辽宁在古生物化石研究、保护及博物馆建设取得的成绩。2012年近5万法国观众参观了于法国奥尔良自然史博物馆举办的"恐龙之声特展",法国国会议员及中国驻法国大使馆公使衔参赞等都参与了此次活动。

2014年11月,博物馆在德国斯图加特自然史博物馆联合成功举办了"中国带羽毛恐龙特展"。开幕式期间,孙革馆长应邀做了有关展览的特邀学术报告并出席记者招待会,斯图加特自然史博物馆馆长、国际古植物学会主席怡德教授发表了热情洋溢的讲话,高度评价中国在"带羽毛恐龙"等研究取得的成就。由于斯图加特市是德国巴登-符腾堡州首

在法国奥尔良联合举办"恐龙之声特展"(2012)
1. 奥尔良市副市长(中)举行欢迎仪式;2. 法方宣传海报;
3. 在奥尔良自然史博物馆与馆长吉野(左3)合影;4. 法国观众观看特展。

与德国斯图加特博物馆联合举办"中国带羽毛恐龙特展"

1. 德国巴登-符腾堡州文化教育代表团来馆访问，沈阳师范大学党委副书记贾玉明（右7）接待代表团，前排就坐的包括10位德国大学校长（2015.9）；2、3. 展厅及观众（2014）。

府，此次特展为2015年巴登-符腾堡州与辽宁省开展科学文化合作发挥了促进作用。

2015年4月起博物馆在法国南特自然史博物馆联合举办的"辽宁带羽毛恐龙特展"取得新的成功。4月8日举行的特展开幕式上，法国最年轻的女市长、南特市市长罗兰和中国驻法国大使馆文化参赞，以及来自法国、比利时等国博物馆、大学的科学家及来宾等百余人出席，南特市报纸及电视台纷纷发表新闻对特展予以高度评价。由于南特市是法国西部滨海城市，许多商业界人士将展览与未来与中国辽宁的经济合作联系起来。此次展览不仅宣传了我国在古生物研究取得的成就、也大大增进了中法之间的科学文化合作和友谊。12月9日，孙革馆长应邀在该馆做了专题科普讲座，法国电视台记者鲍丝（E. Baus）还播放了她新创作的、有关中国热河生物群的科教新片。

法国南特"辽宁带羽毛恐龙特展"(2015)
1. 法国广告及南特报纸新闻；2. 南特市长罗兰(中)在开幕式上致辞；
3. 吉野馆长(右1)、记者鲍丝(左2)及中国大使馆公使衔参赞(左1)参观展览；
4. 吉野馆长(左1)亲自讲解中国化石；5,6. 法国观众参观。

2.6.4 协助建馆

辽宁古生物博物馆的建立,仿佛织出了一条条色彩斑斓的"纽带",将博物馆建设方面的合作推向国际。一个生动的实例表现在日本城西大学邀请辽宁古生物博物馆协助建设城西大学水田纪念博物馆。2011年辽宁古生物博物馆建成后,在沈阳师范大学领导安排下,博物馆热情接待了由著名教育家、日本城西大学理事长水田宗子教授率领的代表团。由于水田理事长的朋友、日本东京大学分子生物学教授大石道夫(M. Oishi)是已故国际著名古植物学家大石三郎(S. Oishi)的长子,古生物世家的熏陶使他酷爱化石,他还从巴西等地收集了众多的鱼化石等标本,这使得水田早就对化石有所喜爱。水田理事长来中国参观了辽宁古生物博物馆后,深深被这座宏伟的博物馆所震撼,促使她萌生了要在自己的大学建立一个博物馆的计划。回国后,她亲自找大石道夫教授协商,又邀请孙革馆长在访日期间与大石见面。大石教授全力支持,辽宁古生物博物馆也在展品(模型)提供等方面予以了大力协助。2013年4月,一座精美的"水田纪念博物馆"终于在位于日本东京市中心的城西大学建成,孙革率代表团应邀出席了开馆庆典,受到热烈欢迎。辽宁古生物博物馆协助日本在大学建设博物馆的工作,也促进了沈阳师范大学与日本城西大学之间的校际合作。

协助日本城西大学建设博物馆
1. 水田理事长等参观博物馆;
2. 沈阳师范大学副校长夏敏(右)在博物馆接待;
3. 孙革馆长在城西大学水田纪念博物馆讲解辽宁化石(模型)。

2.6.5 国内交流与合作

国内交流与合作是辽宁古生物博物馆重要工作之一。建馆五年来,博物馆多次承办国内学术会议,并经常派出馆内工作人员赴兄弟博馆学习等,促进了国内馆际交流与合作。

2013年7月,江苏、河南、安徽、湖北四省古生物学会及江苏省地质学会地层古生物专业委员会与博物馆联合主办了"热河生物群研究前沿学术研讨会",来自全国五省23个单位近百名专家学者以及法国里昂第一大学菲利普(M. Philippe)教授等出席了会议。研讨会召开不仅加深了与会各省自然类博物馆学界的学术交流,也对我国化石保护发挥重要促进作用。

2015年8月11—14日博物馆与中科院南京地质古生物研究所共同承办了"中国古生物学会第28届学术年会"。该年会是本世纪以来中国古生物学会首次在我国东北举办的规模最大的全国性学术年会。中国古生物学会杨群理事长、国土资源部地质环境司李继江处长、国家古生物化石专家委员会办公室王丽霞副主任、中国古生物化石保护基金会陶庆法理事长、辽宁省国土资源厅马原副厅长、沈阳师范大学校长助理刘中海等嘉宾,以及来自全国26个省区90个单位的450余名专家学者出

联合主办"热河生物群研究前沿学术研讨会"

中国古生物学会第28届学术年会(2015.8)

1. 全体代表合影；2. 杨群理事长致开幕辞；3. 大会会场；4,5. 专家研讨；6. 本溪牛毛岭野外考察；7. 季强教授(中右)在北票四合屯介绍"中华龙鸟"。

席。参会代表提交论文摘要近300篇，有大会报告6个，分会场报告215个，展版46个。会议就古生物学及其相关研究领域、古生物教学、博物馆与科普，以及古生物化石保护等进行了深入研讨。会议特邀罗哲西、金小赤、孙革、张克信、汪筱林、黄迪颖等6位专家做大会特邀报告，还颁发了第三届"中国古生物学会青年古生物学奖"。会议期间组织了赴朝阳北票及本溪野外考察，以及参观辽宁古生物博物馆等活动。

2.7 人才培养

要建设好一个博物馆，队伍建设和人才培养至关重要。五年多来，在沈阳师范大学、辽宁省国土资源厅的高度重视和共同努力下，博物馆不断引进人才，队伍建设已取得长足而又迅速的进步，目前已拥有一支高水平专业队伍。

博物馆现有的37名专业人员中，有研究人员16人、技术人员21人。研究人员包括教

博物馆工作人员与时任国土资源部部长徐绍史（前排中）等领导合影（2012.9.5）

授5人、副教授4人、讲师5人、助教2人，其中具有博士学位的有14人，其余均具有硕士学位；45岁以下占2/3。研究人员中有美国植物学会通讯会员1人，全国优秀科技工作者1人，国土资源部高层次杰出青年人才1人，辽宁省优秀专家1人，入选辽宁省百千万工程2人等。

国内外名誉教授或客座教授共20人，含院士6人，包括美国科学院迪尔切院士，英国皇家学会会员、伦敦林奈学会主席爱德华院士，俄罗斯科学院阿克米梯耶夫通讯院士，以及中国科学院院士李廷栋、刘嘉麒、周忠和等一批国际著名地质古生物学家；国外客座教授还包括德国孢粉学家阿什拉夫教授及古脊椎动物学家马丁教授，日本恐龙学家东洋一教授等。国内客座教授包括分子古生物学家杨群，恐龙学家董枝明、徐星，古鸟类学家侯连海，微体古生物学家王成源，孢粉学家尚玉珂，古昆虫学家张俊峰，古无脊椎动物学家段吉业、王五力、张立君，古植物学家郑少林、王永栋及王洪山（现在美国佛罗里达自然史博物馆任职）等。上述优秀专家的加盟为辽宁古生物博物馆的科研、科普和教学等工作发挥了重要指导和支撑作用。

总结博物馆在人才队伍建设的工作体会，主要包括：

（1）引进优秀学术带头人。

在沈阳师范大学、辽宁省国土资源厅共同努力和重视下，2008年从吉林大学引进我国著名古生物学家孙革教授作为辽宁古生物博物馆馆长和学术带头人。孙革教授是中国古生物学会副理事长，中国古生物学会古植物学分会名誉理事长，国家古生物化石专家委员会顾问，国土资源部东北亚古生物演化重点实验室主任，辽宁省古生物化石专家委员会副主任，美国植物学会通讯会员，美国佛罗里达大学（自然史博物馆）名誉教授，日本城西大

博物馆馆长孙革教授

1.孙革教授(中)在新疆野外(2016);2.孙革(中)与前校长赵大宇(右)等在博物馆建设工地(2010)。

博物馆馆长孙革教授获奖证书与奖章(2014)

学顾问,日本福井恐龙博物馆客座研究员,《世界地质》主编;曾任第6届国际古植物学会(IOP)副主席、中国古生物学会秘书长、中国科学院南京地质古生物研究所副所长,吉林大学古生物研究中心主任等;曾获教育部自然科学一等奖、李四光地质科学奖及辽宁省科学技术一等奖等,2014年荣获"全国优秀科技工作者"称号。他不仅在学术上有高深造诣,也是一位富有管理经验的"复合型"专家,并活跃在国际学术舞台。孙革教授的引进,为辽宁古生物博物馆的建设及工作的顺利运行奠定了重要领导基础。

(2)支持年轻专业人才成长。

在沈阳师范大学的努力下,博物馆先后引进了一批优秀的中青年专家,包括胡东宇教授(日本北海道大学博士)、周长付教授(北京大学博士)、段冶教授(澳大利亚迪肯大学博士)、曹成润教授(吉林大学博士)、张宜教授(中科院南京地质古生物研究所博士);副教授刘玉双(首都师范大学博士)、李莉(东北大学博士)及田宁(中科院南京地质古生物研究所博士);以及杨涛(吉林大学博士)、赵鑫(中科院南京地质古生物研究所博士)、陆露(中国

胡东宇教授接受杨群理事长（右图左）颁发的中国古生物学会理事聘书（2013）

周长付教授获中国古生物学会青年古生物学奖（2013）

博物馆年轻专家

张宜教授　　周长付教授　　田宁副教授　　刘玉双副教授　　李莉副教授

杨涛博士　　赵鑫博士　　赵明胜博士　　陆露博士　　梁飞博士　　侯世林博士

地质科学院博士)、赵明胜(成都理工大学博士)、梁飞(吉林大学博士)以及王丽副教授(辽宁大学硕士)等;青年学者侯世林已获东北大学博士,还有2人正在吉林大学攻读博士学位。

胡东宇教授是古鸟类学专家,现任副馆长兼古生物学院副院长,国家古生物化石专家委员会委员、中国古生物学会理事,辽宁省优秀专家,辽宁省古生物演化与古环境变迁重点实验室副主任。他带领课题组首次发现迄今最早的带羽毛恐龙赫氏近鸟龙,其成果2009年在英国《自然》杂志发表,入选"2009年中国高校十大科技进展"和"2009年中国/世界十大科技进展新闻",为我国古鸟类及鸟类起源研究作出重要贡献。

周长付教授是年轻的古脊椎动物学专家,2012年入选国土资源部第一批"国土资源杰出青年科技人才培养计划"和"辽宁省百千万人才工程"(千人层次),2013年荣获"中国古生物学会青年古生物学奖"。他带领课题组首次发现迄今最早的带毛发的侏罗纪哺乳形动物巨齿兽,其成果2013年在英国《自然》杂志发表,他还在龟类及翼龙类化石研究中取得突出成绩,为我国古哺乳动物及古两栖爬行类研究作出重要贡献。

此外,年轻的古植物学专家张宜及田宁、古昆虫学专家刘玉双、古鸟类学专家李莉等也分别在各自研究领域的工作中均取得可喜成绩,其成果均发表在国内外重要学术刊物上;优秀的年轻后备力量杨涛、赵鑫、赵明胜、陆露、梁飞及侯世林等博士以及在职博士生刘晓庆等在工作中也取得可喜成绩。

(3)发挥客座教授作用。

博物馆高度重视客座教授的指导作用。博物馆现有20位名誉教授或客座教授,他们为博物馆的建设、人才培养和国际合作等作出了重要贡献。特别是,我国著名古鸟类学家侯

翼龙中的"素食者"——中国翼龙

古神翼龙科是翼龙大家族中一个奇特的分支,也是一类罕见的"素食者"。它们具有鸟一样的喙,喙短而向下倾斜,类似鹦鹉。故此,古生物学家推测它们可能与鹦鹉相似,也是以森林中的果实和种子为食。在取食过程中,它们可能利用头前部发育的硕大嵴冠,拨开遮挡果实的树枝叶。尽管如此,它们可能也会像现在的素食鸟类一样,时而改善它们的食谱,吃一些虫子和其他小动物。在我国热河生物群中发现的中国翼龙,也是"素食"家族中的一员。不过,中国翼龙的嵴冠较小,或许会更喜欢吃种子。

中国翼龙

作出重要贡献的客座教授

侯连海

徐星

董枝明

张立君

段吉业

傅仁义

郑少林

王成源

张俊峰

尚玉珂

王永栋

王洪山

连海教授、恐龙学家董枝明教授和徐星教授，以及微体古生物学家张立君教授等，为辽宁古生物博物馆的筹建、科研以及人才培养等付出了大量的心血，作出了卓有成效的努力。

客座教授中，古生物学家段吉业、傅仁义、郑少林、王成源、张俊峰、尚玉珂、王永栋、王洪山等也为博物馆顺利运行以及人才培养等做了大量的指导和协助工作。

五年来，博物馆的国内外名誉教授或客座教授的聘任，得到沈阳师范大学领导的高度重视与支持。学校为建设高水平的博物馆，聘任了美国迪尔切院士、英国爱德华院士、俄罗斯阿克米梯耶夫通讯院士，以及中国科学院院士李廷栋、刘嘉麒、周忠和等一批国际著名地质古生物学家为名誉教授；国外客座教授还包括德国阿什拉夫教授、马丁教授及日本东洋一教授等。

（4）重视技术队伍建设。

技术人员队伍是博物馆人才队伍建设中的重要组成部分，

沈阳师范大学领导向院士颁发名誉教授聘书
1. 林群校长向迪尔切院士(左4)、阿克米梯耶夫院士(右3)及阿什拉夫教授(左3)颁发聘书；2. 林群校长向李廷栋院士颁发聘书；
3. 张辉副校长向爱德华院士颁发聘书。

博物馆的技术人员
1. 技术人员王晶琦(左1)赴德国斯图加特自然史博物馆进修；
2. 技术人员白冰、王文帅及张晓巍等参加访问德国森肯堡博物馆；
3,4. 电镜室(3)和设计印刷室(4)的技术人员；5. 技术人员与专家一起赴新疆鄯善恐龙发掘；
6. 展览部讲解员与美国迪尔切院士在一起。

辽宁古生物博物馆对此一直高度重视。博物馆现有21名技术人员分属化石修复与采掘部、实验室、展陈设计室、展览部、电镜室、化石收藏室及信息技术室等7个技术部门,他们绝大多数具有硕士学位,90%以上为40岁以下,主要来自沈阳师范大学生命学院、美术学院和信息技术学院。由于技术部门专业性强,博物馆高度重视对技术人员的培训,并创造条件让他们到校外甚至国外进修学习,每次出国化石展览均有技术人员参加;博物馆还举办"星期四课堂"及专家讲座等,培训他们除具有较高的专业技术水平外,又有较好的地质古生物知识基础。五年来,博物馆技术人员在化石发掘、展览、科普及配合研究人员开展科研工作等方面均发挥了重要作用。

2.8 实验室建设

实验室建设是博物馆成功运行的重要组成部分。五年来,辽宁古生物博物馆在沈阳师范大学和辽宁省国土厅共同支持下,主要从硬件建设(实验室仪器设备)和平台建设(重点实验室)两方面开展工作并取得成效。目前已建设起设备先进的实验室10个,另有省部级重点实验室平台2个。

博物馆实验室建设(1)
1.实验室外景(一层);2.地质标本室;3.孢粉实验室;4.古植物实验室。

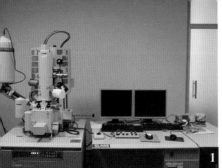
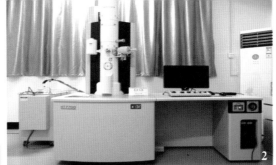

博物馆实验室建设(2)
1. S-4800扫描电镜(SEM); 2. HT-7700透射电镜(TEM); 3. 磨片室实习指导;
4. VHX-600E超景深三维立体显微镜; 5. 设计与印刷室; 6. 化石修复室。

2.8.1 硬件建设

博物馆现有的实验室包括古植物实验室、孢粉实验室、微体实验室、磨片室、化石修复室、电镜室、地质标本室、展览设计与印刷室以及教学实验室(包括古生物、矿物岩石及构造地质的综合实验室),另有由德国孢粉学家阿什拉夫提供显微镜的阿什拉夫实验室1个。实验室总面积约1500平方米,仪器设备总价值约1145万元,上述实验室的建立有力地保证了博物馆科研和相关工作的需要。

电镜室有先进的扫描电镜(SEM,S-4800,场发射),是当今我国最先进的扫描电镜之一,主要用于古生物化石等表面微细观察,其能谱仪在分析物质成分方面也能发挥重要作用。还有透射电镜(TEM,日产HT-7700),主要用于对微体化石的解剖观察等。上述两个电子显微镜的配合,能全方位完成对化石等物体的形态解剖等观察与分析,在分类等研究中发挥重要作用。光学显微镜包括先进的超景深三维立体显微镜(VHX-600E),该显微镜以其特有的功能,可以在不做实验情况下,直接观察到细胞等微观特征。2011—2013年,该显微镜在研究侏罗纪燕辽杉的表皮构造中发挥了独特作用,研究生谭笑在张宜教授指导下取得的成果在德国获国际奖。

2.8.2 平台建设

省部级重点实验室建设是博物馆实验室平台建设的主要组成部分。在辽宁省国土资源厅和沈阳师范大学的共同支持下,五年来,博物馆先后被批准建立了"辽宁省古生物演化及古环境变迁重点实验室"(2012)和"国土资源部东北亚古生物重点实验室"(2015)。此外,博物馆还协助沈阳师范大学创办了我国首家古生物学院(详见4.1),并正在筹建辽宁省"东北亚古生物协同创新中心"等科研创新平台。

"国土资源部东北亚古生物演化重点实验室"是博物馆依托辽宁省国土资源厅,于2012年5月经国土资源部批准建设的,2015年9月正式挂牌运行,旨在通过重点实验室建

国土资源部重点实验室揭牌运行(2015.9)
1. 揭牌仪式,文波副司长(右2)、吴景涛副厅长(左2)及林群校长(右1)等出席;2. 专家评审;
3. 评审专家及领导与重点实验室代表及学生合影,前排左起:张立军、王永栋、万晓樵、罗桂昌、朱友强、孙革、林群、文波、吴景涛、朱立新、郝德永、张辉旭、修玉玲、邵永运、杨建杰。

设,推动博物馆科研、科普和学科内涵建设得到更快发展,真正实现国内一流、国际知名的目标。2015年9月25日,国土资源部重点实验室最终验收及揭牌仪式在博物馆隆重举行。国土资源部科技与国际合作司副司长文波,专家组组长、中国地质科学院常务副院长朱立新,以及万晓樵、罗桂昌、王永栋、朱友强、张立军等专家,国土资源部实验室管理办公室副主任张辉旭,省国土资源厅副厅长吴景涛,沈阳师范大学校长林群、副校长郝德永等出席了验收及揭牌仪式。专家组对实验室所取得的成绩给予了高度评价,一致通过了验收。验收结束后,在博物馆举行了揭牌仪式。文波副司长和吴景涛副厅长在林群校长及孙革主任陪同下,为实验室揭牌。

据悉,以博物馆为单位申请国土资源部重点实验室,在我国国土资源部系统尚属首例。辽宁古生物博物馆被批准正式建立国土资源部重点实验室这一平台建设进展,入选"2015年沈阳高校十大科技新闻"。

2.9 博物馆管理

辽宁古生物博物馆由辽宁省国土资源厅和沈阳师范大学共同领导,辽宁古生物博物馆管理委员会是博物馆的领导和决策机构。管理委员会主任由辽宁省国土资源厅厅长季风岚担任(兼博物馆名誉馆长);副主任由沈阳师范大学校长林群担任。管理委员会委员包括:马原(辽宁省国土资源厅副厅长)、王大超(沈阳师范大学副校长)、孙革(辽宁古生物博物馆馆长)、郭杰(辽宁省化石保护局局长)及杨杰(沈阳师范大学校务办公室主任),共7人组成。五年来,在博物馆管理委员会的直接领导下,博物馆管理工作有序开展,并特别重视制度建设在博物馆管理工作中的重要作用。

2.9.1 博物馆管理委员会

主　任:季风岚(辽宁省国土资源厅厅长、博物馆名誉馆长)

副主任:林　群(沈阳师范大学校长)

委　员:马　原(辽宁省国土资源厅副厅长)

　　　　王大超(沈阳师范大学副校长)

　　　　孙　革(辽宁古生物博物馆馆长)

　　　　郭　杰(辽宁省化石保护局局长)

　　　　杨　杰(沈阳师范大学校务办公室主任)

辽宁古生物博物馆管理委员会成员

季风岚
（主任、厅长、名誉馆长）

林 群
（副主任、校长）

马 原
（副厅长）

王大超
（副校长）

孙 革
（馆长）

郭 杰
（局长）

杨 杰
（校办主任）

2.9.2 制度建设

为做好博物馆管理工作，辽宁古生物博物馆不断加强制度建设。五年来，根据工作的实际需要，针对博物馆的工作职责、人员考核、安全保障、财务管理、实验室及标本库房管理等方面，制定了数十项工作制度，主要包括《古生物博物馆工作职责》（共12项）,《古生物博物馆管理制度》（共33项）等。管理制度包括博物馆藏品管理制度、实验室管理制度、实验室危险品管理制度、电镜室管理制度、切片机及磨片机安全操作规程、资料室书刊借阅管理制度、工作人员考核办法（暂行）、财务工作管理制度、固定资产管理办法、车辆管理与使用制度、档案工作管理制度、外事工作规章制度、安全工作管理制度、突发事件应急预案实施办法等。博物馆向全馆工作人员颁发了《辽宁古生物博物馆工作制度汇编》，使全馆各项管理工作更加规范和科学，促进博物馆依法、有序地运行和发展。

博物馆管理
1. 博物馆入口安检；2. 消防安全检查（左2为馆长助理孙大勇）；
3. 检查监控室设备；4. 博物馆安全科普宣传。

第三章

辽宁化石保护工作

3.1 辽宁化石保护工作

辽宁省因盛产热河生物群和燕辽生物群等珍贵的古生物化石,以及取得众多世界级科研成果而享誉海内外,辽宁也因此成为国际著名的古生物化石产地和重要古生物化石保护区之一。2013年,辽宁的朝阳、义县和建昌三地被国土资源部批准为"国家重要古生物化石集中产地"。

辽宁古生物博物馆宣传辽宁化石保护的展板

首届辽宁省古生物化石鉴定委员会工作会议(2001.1,沈阳)前排(左起)李煜、孙革、侯连海、吴启咸、张鹏发、田启山、董桂福、王欠教
The 1st meeting of Liaoning Committee of Fossil Identification (Shenyang, Jan. 2001)

第二届辽宁省古生物化石鉴定委员会工作会议(2002.9,沈阳)
The 2nd meeting of Liaoning Committee of Fossil Identification (Shenyang, Sept. 2002)

第三届辽宁省古生物化石鉴定委员会工作会议(2006.5,沈阳)前排左起:郑少林、董拥军、郭胜哲、李强、孙伟凤、孙革、侯连海、董枝明、王五为等
The 3rd meeting of Liaoning Committee of Fossil Identification (Shenyang, May 2006)

辽宁省古生物化石保护现场会(辽宁凌源,2002)
The meeting of Fossil Protection in Lingyuan of Liaoning (2002)

近20年来,古生物化石越来越受到重视。但受经济利益等驱使,辽宁西部等一些化石产地频频出现乱采乱挖的现象,一些不法分子倒卖化石、从中渔利,更有甚者将化石偷运出境。辽西许多化石产地"千疮百孔",许多地方的盗采盗挖屡禁不止。

为此,如何保护好古生物化石及其珍贵的化石产地?如何让广大群众自觉懂法、守法?如何打击破坏化石的违法犯罪现象?如何将化石及其产地合理开发利用?已成为各级政府工作面临的挑战,也特别成为主管辽宁化石保护的辽宁省国土资源系统的新课题。

辽宁省的古生物化石保护工作在我国开展最早,20年来通过辽宁各级政府在化石保护工作中的不懈努力,取得了很大成绩,并积累了宝贵的经验。自1997年以来,辽宁在全国首建化石保护区;2001年率先颁布《古生物化石保护条例》,制定了《古生物化石定级标准》,并率先成立辽宁省古生物化石鉴定委员会。辽宁开展的大量化石保护工作为国家制定《古生物化石保护条例》及其实施办法等奠定了重要基础。由此,辽宁也被誉为我国"古生物化石保护的一面旗帜"。

总结辽宁在化石保护工作取得的主要工作经验,似可归纳为如下几个方面。

(1)开展化石及产地调查,"摸清家底"。

辽宁省国土资源厅为做好化石保护工作,首先做到"摸清家底"。截至2015年,他们已在全省化石资源调查中确定古生物化石点247处,其中重点化石点165处;完成了《辽宁省化石资源调查评价报告》、《辽宁省古生物化石点名录》和《辽宁省化石资源分布图》等;并于2014年在喀左和建昌两县完成了1∶10万古生物地质调查,对重点区进行了更详细的地质填图工作,为化石保护工作奠定了重要基础。

(2)建立保护区和地质公园,加强化石保护。

1997年,原辽宁省地质矿产厅率先建立了北票鸟化石群自然保护区,1998年该保护区经国务院批准升格为国家级保护区。此后,辽宁省国土资源厅又先后在义县(2002)、朝阳(2003)、本溪(2006)等地建立保护区或地质公园。2004年,经国土资源部批准,建立朝阳鸟化石群国家地质公园,2008年、2010年又先后在建昌和义县成立了保护区或地质公园。2014年,义县、朝阳和本溪三个化石产地均被国土资源部确定为"首批国家重点古生物化石集中产地",大大加强了全省化石的保护力度。

(3)制定法律法规,使化石保护有法可依。

2001年,原辽宁省地质矿产厅率先制定了《辽宁省古生物化石保护条例》,并制定了《辽宁省古生物化石定级标准》,该条例曾于2001年1月经辽宁省人大通过,为后来国家制定《古生物化石保护条例》(2010)和《国家古生物化石分级标准(试行)》(2012)等发挥了重要参考作用。此后,辽宁省国土资源厅又制定了《省级古生物化石保护规划编制指南》

（2012）等文件,发表了一批有关化石保护的文章[50,54,63],进一步推动了辽宁省化石保护的管理。目前,辽宁省国土资源厅正在着手制定新一轮化石保护规划,使辽宁的化石保护工作更上一层楼。

（4）紧密依靠专家指导。

紧密依靠专家指导、提高化石保护工作的科学性,是辽宁化石保护工作的一大特色。2001年1月,原辽宁省地质矿产厅（现国土资源厅）在沈阳主持召开了"全省化石资源保护管理工作会议",并率先成立"辽宁省古生物化石鉴定委员会",邀请了著名古生物学家董枝明、侯连海、孙革、季强、王元青、姬书安、张海春、南润善、郭胜哲、郑少林、张立君、巩恩普等参加会议,形成了我国省级地质管理部门实力最强、水平最高的古生物专家咨询机构。鉴定委员会多次召开会议,为辽宁省化石保护法律法规制定、化石分类级别认定及化石保护管理等发挥了重要作用。

2014年,辽宁省国土资源厅又成立了《辽宁省古生物化石专家委员会》,由时任副厅长杨德军任主任,古生物学家、辽宁古生物博物馆馆长孙革和时任化石保护局局长孙永山（现国土资源厅副巡视员）任副主任,古生物学专家董枝明、侯连海、季强、王元青、郭胜哲、

辽宁省古生物化石鉴定委员会第一、第二次会议

1. 专家研究化石定级,左起:姬书安、王元青、季强、董枝明、吴启成、侯连海、孙革、王丽霞;
2. 首次会议（2001）,前排左起:季强、孙革、侯连海、吴启成、张殿双、田连山、董枝明、王元青;
3. 第二次会议（2002）,左列（左起）季强、孙革、董枝明,右列（右起）赵义宾、张殿双、王丽霞、南润善、傅仁义等。

辽宁省古生物化石鉴定委员会第三次会议（2005）
前排左起：郑少林，南润善，郭胜哲，季强，季风岚，孙革，侯连海，董枝明，王五力，张建平。

郑少林、张立君、王五力、孙春林、巩恩普、汪筱林、任东、吕君昌、王永栋、傅仁义、高春玲、杨雅君，以及化石管理专家王丽霞、张焕翘、张立军等共30余人担任委员。该专家委员会的成立为辽宁省化石保护工作注入了新的活力。

（5）加强博物馆建设，促进化石保护。

为加强化石保护和科学普及等工作，进一步贯彻落实国家《古生物化石保护条例》指示精神，近年来，辽宁省政府大力推进地质古生物博物馆建设，取得可喜成绩。由于政府的重视和合理引导，一批批新的地质古生物博物馆陆续建立。各级博物馆的建立，不仅有利于化石保护与管理，也有利于将化石保护与科研及科普等进一步密切结合。截至2015年12月，辽宁各地建立的古生物（或地质古生物）博物馆有12座，正在建设中的有4座，辽宁已成为全国公立及民间古生物化石博物馆成立最多的省份。辽宁省国土资源厅与沈阳师范大学共建的辽宁古生物博物馆也是辽宁化石保护工作的标志性成果之一。

辽宁省化石保护会议及制定化石保护文件
1. 沈阳会议（2001）；2. 凌源会议（2002）；
3. 辽宁省国土资源厅颁发的部分文件。

辽宁省古生物化石专家委员会成立
前排左起：郭胜哲，傅仁义，张立君，孙永山，孙春林，李强，王丽霞，杨德军，孙革，王元青，
巩恩普，王永栋，张焕翘，王五力，郑少林。

3.2 博物馆共建

为贯彻落实国家有关化石保护的指示精神，创造性地做好辽宁的古生物化石保护工作，辽宁省国土资源厅自2005年起，果断地作出了与沈阳师范大学共建"辽宁古生物博物馆"的决策。为了切实建好这一我国规模最大的古生物博物馆，辽宁省国土资源厅从政策支持、资金提供以及施工指导等诸多方面都作出了极大的努力。博物馆建设期间，省国土资源厅季风岚厅长等领导曾多次来施工现场视察博物馆建设情况，并与沈阳师范大学领导共商建设大计。在省国土资源厅和沈阳师范大学密切协作下，自2006年6月6日正式动工起，历经近五年的艰苦努力，一座雄伟壮观、具有国内外一流水平的古生物博物馆终于拔地而起，于2011年5月与世人见面。

目前辽宁古生物博物馆已成为集展示、收藏、科研、科普及教学功能五位一体的优秀的自然类博

辽宁省国土资源厅领导视察博物馆建设工作
1. 季风岚厅长（左3）等听取沈阳师范大学领导关于博物馆建设情况汇报（右2为前校长赵大宇）；
2,3. 季风岚厅长视察展厅筹备。

物馆之一,已成为我国古生物科研、科普和教学的主要中心之一以及辽宁省古生物科研与科普的核心基地,在辽宁省化石保护工作中发挥重要作用。辽宁古生物博物馆的建立是辽宁省国土资源厅在化石保护管理工作中的一个创举。正如时任国土资源部部长、现任国家发改委主任徐绍史所指出的:辽宁古生物博物馆的建立是政府与高校合作共建博物馆成功的典范。

徐绍史、汪民等领导来博物馆视察
1. 徐绍史主任(左3)参观恐龙化石,陈超英副省长(左2)、关凤峻司长(右2)、于文明书记(左1)、林群校长(后排中)等陪同;2. 汪民副部长(左3)等领导参观博物馆化石;
3. 徐绍史主任等领导在博物馆前合影,左起:郝德永、贾玉明、季凤岚、林群、徐绍史、陈超英、于文明、孙革、王大超。

第四章
辽宁建立古生物学院

走进辽宁的古生物世界,让我们看到了辽宁30亿年来的化石和它们漫长的演变历程,看到了收藏、研究和展示数以万计化石的辽宁古生物博物馆,也了解了辽宁化石保护工作的艰辛和成绩。但辽宁的化石研究、科普和保护工作未来由谁来做?博物馆的古生物学及其相关专业人才未来从哪里来?我国大学是否还应有古生物专业?这些新的问题又摆在面前。

4.1 古生物人才培养的新摇篮

辽宁古生物博物馆的成立为辽宁古生物化石保护、科研和科普等工作的开展带来了新的希望。但博物馆如何做好古生物学人才的培养?作为综合性大学的沈阳师范大学,能否设置古生物专业?沈阳师范大学可以培养古生物学研究生,但古生物专业的本科生又从哪里来?这些问题又提出了新的挑战。

我国高校古生物专业设置已有近百年历史,最早是20世纪20年代由李四光院士等老一辈古生物学家于北京大学创建。当时,自英国伯明翰大学留学回国的李四光教授和美国葛利普教授在北大执教,我国后来的古生物学家杨钟健、俞建章、乐森璕、斯行健、赵金科、卢衍豪等都从这里毕业,也都曾是李四光先生的门生。

中华人民共和国成立后,北京大学、南京大学、北京地质学院(现中国地质大学)、长春地质学院(现吉林大学)等高校都设有古生物专业(或专门化),乐森璕、俞建章、王鸿祯等老一辈古生物学教育家为此曾付出毕生的心血[68]。

但20世纪60年代后期,由于"文革"等干扰,加之地质行业不景气等,我国高校的古生物专业相继被取消,以至招收的古生物研究生的专业背景五花八门,这给古生物学优秀人才培养带来许多困难。而近年来,随着国内科研、石油、地质以及博物馆等部门对古生物专业人才的需求日益增长,特别是中国科学院南京地质古生物研究所和古脊椎动物与古人类研究所等我国古生物学教育与研究的主要机构,也迫切需要有高质量的古生物专业

我国老一辈古生物学家和古生物教育与研究主要机构
1. 李四光院士；2. 李四光和斯行健院士（右）抗战时期在桂林；3. 原长春地质学院副院长俞建章院士；
4. 北京大学——中国古生物学家的摇篮，1920年李四光在这里创办古生物专业；
5. 中科院南京地质古生物研究所；6. 中科院古脊椎动物与古人类研究所。

本科生作为研究生的生源。事实表明，目前急需在我国、至少应在部分有培养能力的高校尽快设立（或恢复）古生物专业。

　　于是，根据目前社会对古生物专业人才的需求，结合辽宁古生物博物馆（研究所）具有一支高水平的师资队伍、沈阳师范大学还有生命学院可作为教学依托，以及辽宁古生物博物馆可作为一个很好的教学实习平台等情况，我国古生物学家、辽宁古生物博物馆馆长孙革率先提出，申请在沈阳师范大学建立我国首家以学院为建制的古生物学院，此举得到了学校领导的全力支持。2010年12月，沈阳师范大学正式批准古生物学院成立，由博物馆馆长孙革教授兼任院长，2011年9月正式迎来首批古生物专业本科生入学。

　　2011年10月9日，沈阳师范大学校园洋溢着喜庆的气氛，沈阳师范大学古生物学院成立仪式在辽宁古生物博物馆隆重举行。中国古生物化石保护基金会前理事长、国土资源部前副部长蒋承松及副理事长张卫东，中国科学院院士李廷栋，国家古生物化石专家委员会办公室副主任王丽霞，中国古生物学会秘书长王永栋，北京大学深圳研究生院院长白志强，吉林大学古生物研究中心主任孙春林，辽宁省国土资源厅前副厅长杨德军，沈阳师范大学校长林群、副校长夏敏以及古生物学院师生等近百人出席了成立仪式。仪式上，林群校长代表学校首先由衷感谢社会各界的大力支持；他指出，古生物学院的成立是沈阳师范大学为学校古生物学科建设和人才培养、为我国古生物学事业发展和辽宁省的科技教育事业做出的新举措，希望古生物学院能不断产出高水平科研成果，培养出高素质的优秀学生成为中国古生物事业优秀的接班人。蒋承菘、李廷栋、杨德军等嘉宾和林群校长共同为古生物学院揭牌，中国古生物化石保护基金会和辽宁朝阳文化传奇有限公司分别为古生物学院捐赠了奖学金。仪式结束后，与会领导和嘉宾参观了古生物学院和博物馆，沈阳师范大学党委书记于文明举行了接待会并致辞，他除代表学校向来宾们的大力支持表示感

古生物学院成立仪式（2011.10）

1. 林群校长（左1）、李廷栋院士（左2）、蒋承菘前副部长（右2）和杨德军副厅长（右1）为学院揭牌；
2. 张卫东副理事长（左）向古生物学院赠送奖学金；
3. 于文明书记在接待会上致辞；
4. 于书记（左4）与来宾及学院教师在接待会上；
5. 全体合影。

古生物学院
1. 学院教学楼；2. 名誉院长、美国迪尔切院士讲课；
3. 院长孙革教授亲自上电镜课；
4. 学院领导（左起）：王亚君（副书记），胡东宇（副院长），孙革（院长），杨建杰（书记），曹成润（前副院长），张洪刚（副院长）。

谢外，还希望古生物学院今后能够为古生物学研究和国家化石保护工作等输送优秀的专业人才，为我国及国际古生物学科发展增添"后劲"，也为辽宁的化石保护工作多作贡献。

经过五年的运行，目前，沈阳师范大学古生物学院已成为我国高校在古生物学本科教学和人才培养方面的一支新的"生力军"，也是目前国际上唯一的一所古生物学院。学院下设古生物学、化石保护与博物馆学、化石能源学等3个方向，每年招收本科生约30人，研究生5人；现有本科生113人、研究生15人。学院有一支高水平的教师队伍，现有20余名教师中，有教授5人，副教授4人，讲师6人，具博士学位的有16人，均毕业于国内外知名高校或研究机构。2011年底经学校批准，聘请了美国科学院迪尔切院士为古生物学院名誉院长，进一步扩大了学院的国际影响。

五年来，学院在科研、教学、国际交流与合作等方面均取得突出成绩；在英国《自然》杂志发表论文6篇（含第一作者3篇），SCI论文50余篇，出版专著及教材8部；获国家科技专项及国家自然科学基金等8项，获省部级一等奖1项、二等奖1

古生物学院教学与实习
1. 化石角质层分析实验课；
2. 磨片实验课；3. 新疆野外实习；
4. 俄罗斯野外实习。

项。目前,古生物学院的古生物学专业已成为沈阳师范大学的支柱性和标志性专业之一,古生物学院的迅速发展也为沈阳师范大学的学科建设和专业建设发挥了积极促进作用,在国内外产生广泛影响。

在教学工作中,古生物学院突出"科研带动教学、重视实践教学、国际化"等特色,毕业生普遍具有"基础功底扎实、动手能力强"的特点。建院以来,除课堂上学习专业知识外,学生们在小学期教学实践活动中,先后赴本溪、秦皇岛和新疆等地野外实习,还掌握了化石角质层分析、孢粉分析及化石磨片等实际操作技能。2016年夏,学院又在黑龙江嘉荫和新疆鄯善先后建立新的教学实习基地。2014年7月,2011级本科生赴俄罗斯远东阿穆尔国立大学和结雅—布列亚等地区野外实习,大大开阔了学生们的国际化视野,增强了对外交流能力。

研究生谭笑2013年在德国获国际奖

春华秋实,硕果满园。五年来,沈阳师范大学古生物学院已有两届本科生和三届研究生毕业;2011级和2012级本科生共23名考取了研究生(考研率43%),包括英国格拉斯哥大学、北京大学、中国地质大学(北京、武汉)、中科院南京地质古生物研究所、中科院古脊椎动物与古人类研究所等著名高校或研究机构;4名硕士研究生考取中科院古

本科生在"辽宁太阳能杯"大学生辩论赛获奖

优秀毕业生

裘锐	刘美彤	刘璐	谢奥伟	王英杰
中科院古脊椎动物与古人类研究所博士生	英国格拉斯哥大学硕士生	北京大学硕士生	中科院南京地质古生物研究所硕士生	上海睿宏文化传媒公司

2012级毕业生合影(2016)

脊椎动物与古人类研究所、东北大学和吉林大学的博士生。硕士生谭笑2013年在德国格丁根国际会议上获奖;本科生王蕾2014年以第一作者在国内核心期刊发表论文,获辽宁省高校大创项目一等奖。

随着国家自然类博物馆和文化产业部门用人需求的增加,2011级和2012级毕业生共有18名同学毕业后分别被重庆自然博物馆、安徽地质博物馆、常州恐龙园博物馆、大连星海古生物博物馆、上海睿宏文化传媒公司、北京天图文化产业公司和常州卓谨文化传媒公司等招聘录用,很快在实际工作岗位上发挥作用;毕业生王英杰、刘思麟、李青蒂等都受到用人单位的好评。

4.2 学院与博物馆

辽宁古生物博物馆和沈阳师范大学古生物学院像是一对"孪生兄弟",他们在工作中密切配合,共同见证了古生物学科研和科普等工作在辽宁的快速发展,也分享着在古生物学人才培养方面捷报频传的喜悦。

古生物学院的学生(包括研究生)和教师的人力资源和智力资源是辽宁古生物博物馆的宝贵财富。科普及培训方面,学院为博物馆提供了有力的支撑;而古生物学院的实践教学和社会服务实践主要在博物馆开展。五年来,古生物学院的师生积极参加辽宁古生物博物馆的科普活动,学院全体学生都是博物馆的"志愿者",在博物馆科普中发挥了重要作用。李莉副教授亲自为"小小讲解员"培训授课,田宁副教授主动赴彰武县的边远山区学校开展"科普进乡村"活动,都深受广大群众特别是青少年的喜爱。

除科普活动外,学院老师们还开展了"星期四课堂"活动,利用博物馆每周四业务学习时间给博物馆工作人员(特别是讲解员和技术人员)授课,并介绍国内外先进的自然类博物馆的经验,以"国际化"的标准提出提升业务水平的希望等。由此,博物馆除展示、收藏、

古生物学院师生积极参加博物馆科普
1. 学生参加招募志愿者活动；
2. 研究生裴锐为孩子们讲授恐龙知识；
3. 田宁副教授在辽宁彰武县农村学校做科普报告；
4. 学院副书记王亚君主持博物馆志愿者大会。

科研、科普功能外，又增加了教学功能，一个"五位一体"的、坚实的科学文化教育平台正在辽宁古生物博物馆形成。

学院与博物馆共建和紧密融合，一方面可以利用高校教学与科研等资源为博物馆建设服务，另一方面，博物馆也可以成为高校师生为社会开展文化科普等服务的一个"实验田"和"窗口"，院馆的密切合作也可能在未来会闯出一条"协同创新"的新路。

辽宁古生物博物馆，作为我国自然类博物馆大家庭中一个刚满5周岁的"娃娃"，正像一颗冉冉升起的新星，在科学家和社会大家庭的热情关怀和呵护下，不断增加新的光彩。希望它能在国际古生物学和博物馆学浩瀚的太空中，逐渐闪烁出它特有的光辉。

辽宁古生物博物馆部分成员合影(2011.5)

致 谢

我国唐代诗人贾岛的名言"十年磨一剑"喻示着为实现理想而聚多年努力、精心完成宏大事业的精神。辽宁古生物博物馆经历了前五年的辛勤建设和后五年的成功运行,前后也整整经过了十年的历练。纵观十年来辽宁古生物博物馆走过的历程,作者们由衷感谢辽宁省政府、特别是辽宁省国土资源厅和沈阳师范大学,感谢他们为国家古生物化石保护和学科发展所作出的英明决策和为此而付出的卓有成效的努力和实干精神。

特别要感谢辽宁古生物博物馆建设的功勋领导者——沈阳师范大学前校长赵大宇教授。他为博物馆的建设蓝图设计和施工、为博物馆的运行策划和人才引进等,都倾注了大量心血。他在管理上超人的胆略、务实的作风以及广博的知识底蕴等,促进了将辽宁古生物博物馆建设成集科学、文化、艺术为一体的博大精品。今天,每当我们走近这一雄伟高耸的博物馆时,我们仿佛看到一位曾在风雨中指挥博物馆建设的身影。赵大宇教授对辽宁古生物博物馆建设的杰出贡献,人们将永志不忘。

辽宁古生物博物馆建设的功勋领导者——赵大宇
1. 沈阳师范大学前校长赵大宇;
2. 赵大宇向国土资源部前副部长寿嘉华(左2)等介绍辽宁古生物博物馆模型(2010)。

沈阳师范大学为辽宁古生物博物馆主办MTE-12国际会议文娱演出

沈阳师范大学梅兰芳研究所所长、国家一级演员肖迪(后排右9)率师生为国内外来宾演出(2015.8.16)。

　　对辽宁古生物博物馆建设和运行的成功,还要感谢多年来沈阳市政府和国内外专家以及社会各界的大力支持与帮助。特别感谢中国科学院李廷栋、殷鸿福、刘嘉麒、舒德干、周忠和等院士;感谢国土资源部、中国科协、中国古生物学会、国家古生物化石专家委员会、中国古生物化石保护基金会、中科院南京地质古生物研究所、中科院古脊椎动物与古人类研究所、中科院昆明植物研究所及西双版纳植物园、中国地质科学院、吉林大学、东北大学、中国地质大学(北京、武汉)、北京大学、中山大学、贵州大学、西北大学;以及中国地质博物馆,中国古动物馆,北京、大连、重庆等自然博物馆,自贡、嘉荫等恐龙博物馆,甘肃、河南、安徽、新疆、吉林大学等地质博物馆,常州恐龙园,南京、深圳、朝阳、大连星海、朝阳济赞堂、锦州、义县等古生物博物馆,上海睿宏和常州卓谨等文化传媒公司,《化石》《生命进化》及《大自然》等杂志;辽宁省化石保护管理局,本溪、朝阳(包括北票)、葫芦岛(包括建昌)、锦州等国土资源部门。也衷心感谢沈阳师范大学生命学院、戏剧学院及梅兰芳艺术研究所、音乐学院以及机关各有关部门的多方支持与协助。

　　借此机会,也感谢国外专家多年来的支持与协作,包括美国科学院院士D. L. Dilcher,俄罗斯科学院通讯院士M. Akhmetiev,英国皇家学会院士D. Edwards,德国科学院院士V. Mosbrugger,以及(按英文字首顺序)A. R. Ashraf, Y. Azuma, Yu. Bolotsky, E. Bugdaeva, J. Eder, F. Escuillie, L. Golovneva, P. Guillet, I. Harding, A. Herman, K. Johnson, D. Jones, T.

Kezina,T. Kodrul,V. Markevich,T. Martin,S. Naugolnikh,H. Nishida,M. Popa,I. Pospelov, A. Roisenberg,R. A. Spicer,S. Suzuki,M. Tekleva,K. Terada,E. Volynets,A. Zinkov,N. Zavialova等专家,还有我国在外工作的专家罗哲西教授和王洪山教授等。

与此同时,也特别感谢沈阳方洋文化公司总经理房良、朝阳传奇文化产业公司董事长李海君先生等;他们曾向辽宁古生物博物馆无私馈赠个人手中收藏的珍贵化石,有力地支持了辽宁古生物博物馆的建设和顺利运行。

本书的出版得到国家科技部科技基础性工作专项《中国标准地层建立——中国地层表的完善》(2015FY310110-15)、中国地质调查局《全国陆相地层划分对比及海相地层阶完善》项目(121201102000150010)、教育部与国家外专局"111"项目(B06008)、国土资源部东北亚古生物重点实验室、辽宁省古生物演化与古环境变迁重点实验室、东北亚古生物演化与环境教育部重点实验室、教育部科研专项资助以及沈阳师范大学等经费资助。

本书写作过程中,得到段吉业、张立君、郑少林、张俊峰等古生物学家的大力协助;稿件修改中得到彭善池教授的指导。在本书写作辅助工作中,辽宁古生物博物馆吴思竹协助撰写英文,白冰、胡进、刘森、王文帅、何佳怡、宋佳乐、于鹏等均给予热诚协助。作者在此一并表示衷心的谢意。

本书编辑出版中,得到上海科技教育出版社王世平总编、编辑伍慧玲博士及汤世梁老师等的热诚支持和多方帮助。作者们由衷敬佩他们严谨的工作作风、忘我的工作热情和上乘的编辑水平,感谢他们在本书编辑出版中所付出的智慧和辛勤汗水。

Entering Paleontological World in Liaoning

By
Sun G., Hu D. Y., Zhou C. F., Liu Y. S., Yang T.,
Fu R. Y., Cheng S. L., Yang J. J., Zhang H. G., Sun Y. S., Wang L. X.

Preface

The earth we live experienced about 4.6 billion years (Ga) of evolution, and this evolution is still in process. The earth's atmosphere began to transit to "greenhouse" along with the increase of CO_2 since about 260 million years (Ma) ago. However it began to transit to "ice house" from late Cenozoic about 3 Ma ago. The formation and strengthening of Asian monsoon aggravated the desertification of lithosphere. Today, the pollution of hydrosphere has been increasing, and the diversity of biosphere has been dwindling. All these grim changes of the earth have taken place in front of human beings. For better understanding and improving the effects of the earth's changes, we should get more information about the causes, process and rule of the changes through 4.6 Ga, especially the changes in the earth's biological world and environmental evolution so as to find out some principle and enlightenment. This is not only the job for geoscientists, particularly the paleontologists, but also the expectations of science amateurs.

About 3.8 Ga years ago, bacteria and cyanobacteria, the earliest forms of life, appeared in proto-ocean on the earth. These early forms of life developed from single cells to multi-cellular cells, from simple to complex. The earliest Metazoan appeared in over 600 Ma ago, and the earliest terrestrial vascular plant appeared about 400 Ma ago. From this time, the earth bade adieu to monotony and loneliness, and gradually changed to today's vigorous and colorful world. How do we know the long evolution of life in geological history? What wonderful stories happened during the long evolution process? What suggestions can we make? We get all these information from the fossils of life preserved in rocks. Fossils, the precious nature legacy which recorded billions of years' of sea and land changes and life evolution on the earth, unfolded one by one magic stories they have experienced.

Liaoning, where the earliest rock record in China can date back to 3.8 Ga ago, is one of the oldest regions in the geological history of China, even in East Asia. The earliest fossil record—the primitive cyanobacteria of Anshan Biota—can date back to ca. 2.5 Ga ago. Therefore, walking into geological and paleontological world of Liaoning opens a remarkable "door" of knowing the evolutionary history of geology and life in China. Looking over fossils in Liaoning not only helps science fans to enrich geological and paleontological knowledge, but also lets them have endless reverie cruising in the ocean of knowledge and thinking about the future of Earth and humankind.

Since 2006, there has been a feat in paleontological career in Liaoning, a Fossil Province. Department of Land and Resources of Liaoning (DLRL) and Shenyang Normal University (SNU) began to co-build Paleontological Museum of Liaoning (PMOL). After five years of efforts, the PMOL officially opened on May 21, 2011. The museum, 30 metres tall with 15,000 m^2 building area, has become the biggest paleontological museum in scale in China. The museum shows "the world fossil treasury" in Liaoning to the public. The rare collections of fossils include the Jehol Biota and Yanliao Biota, such as the earliest feathered dinosaur *Anchiornis*, the earliest known mammaliform with hair *Megaconus*, the earliest known flowers *Archaefructus*, the earliest glide lizard *Xianglong*, the biggest known dinosaur in Liaoning, *Liaoningotitan*, and some newly found primitive birds in recent years, *Shenshiornis*, *Bohaiornis*, *Shengjingornis*, and some turtle fossil new taxa, such as *Liaochelys*. The study of these rare fossils has made great contributions to solving many problems related to the origin of birds, the angiosperm origin and early evolution, etc. In the past two decades, many research results on the fossils of Liaoning have been published on *Nature* (UK), *Science* (US) and some other global high-level academic journals, in which some achievements were selected as Top Ten Science & Technology News in China or Top 100 Science News (US). Since the finding of these fossils, Liaoning is called "the place where the first bird flew and the first flower bloomed". Therefore, visiting with introducing the PMOL is significant for China's science and culture, especially for science popularization. This is why we write and publish this book.

This book is the companion volume to *The Fossil Record of Three Billion Years in Liaoning* published in 2011. However this book has a lot of new information, such as the addition of more contents about the "top ten fossil biota in Liaoning", particularly about the Yanliao Biota, Jehol Biota and Fushun Biota. In addition, specific introduction to the PMOL and its fossils in some details are given, including the "ten fossil stars" of the PMOL for the first time; more discussions are included on the important roles of the earliest feathered dinosaur *Anchiornis* and of the earliest known flowering plant *Archaefructus* in the research of the origins of birds and angiosperms, and their new influences in the world. Chapter 2 is a review of the construction process of the museum, especially of the construction and operational ideas of the museum, important scientific research, science popularization, international cooperation, and the preliminary experience and useful trials of the co-construction of museum by government and university. Moreover, the publication of this book is also an expression of appreciation for fossil protection and science popularization in Liaoning.

In a word, this book will guide readers into the three billion years of wonderful paleontological world in Liaoning, and help readers appreciate the paleobiodiversity of Liaoning, and understand the new development of China's geological and paleontological museum career and the new concept that aims at "internationalization" from a new aspect. Readers can not only get edification about paleontological knowledge, but also some enlightenment about the museum construction from this book.

May 21, 2016, is the fifth anniversary of the PMOL. The book is a present for its birthday. The authors sincerely thank years' of support from the Ministry of Land and Resources of China (MLRC), the DLRL and BFPL (Bureau of Fossil Protection of Liaoning), and the SNU. Also the authors appreciate the assistance from colleagues of the PMOL and CP-SNU (College of Paleontology, Shenyang Normal University), especially Professors Duan J. Y., Zheng S. L., Zhang L. J. and Zhang Y. for their contributions to the book.

The publication of the book is attributed to the support from Key-Lab of Evolution of Past Life in NE Asia (MLRC), Key-Lab of Evolution of Past Life and Environment in NE Asia (MOEC), Key-Lab of Evolution of Past Life and Paleoenvironmental Changes of Liaoning Province, and the 111 Project of China. Thanks should also be extended to the Editor-in-Chief Wang S. P., and editors of Wu H. L. and Tang S. L. from Shanghai Scientific & Technological Education Publishing House for their great support in editing and publishing work.

Prof. Dr. Sun G.
August, 2016 in Shenyang

Chapter 1
Entering Paleontological World in Liaoning

1.1 Geological changes in the past three billion years

Fossils are remains, including organisms body and trace, which lived on the earth in the geological time. Located in the eastern part of the North China tectonic plate, Liaoning is one of the oldest areas in the geological history of China. The record of the earliest known Archean granitic rocks in Anshan of Liaoning dates back to ancient times, ca. 3.84 Ga ago. In the Archean Upper Anshan Group composed mainly of metamorphic rocks (ca. 2.5 Ga), the primitive cyanobacteria and iron bacteria were found.

Between the Late Archean and Early Proterozoic in Liaoning, cyanobacteria and iron bacteria released oxygen in seawater, which let the divalent iron ions (Fe^{2+}) in the water absorb oxygen and change into the ferric (Fe^{3+}) precipitation; and prompted Fe_2O_3 to increase and form hematite, and even further magnetite. The formation of the large quantity of iron deposits and corresponding minerals output in the Late Archean Anshan Group and the Early Proterozoic Liaohe Group is one of the important events in the geological history of Liaoning.

About 25–16 Ga ago, the Early Proterozoic strata were mainly distributed in southern Liaoning, represented by the Liaohe Group. From Mid-Proterozoic to Early Neoproterozoic, the Liaoning area was covered by the "Yanliao Sea", where the deposits were over 10,000 m in thickness, and distributed mainly in western Liaoning and subdivided (in ascending order) into the Changchengian- and Jixian Systems (1.6–1.0 Ga for both), and the Qingbaikouan System (1.0–0.8 Ga). During the time, the fossils were mainly stromatolites. The Middle-Late Neoproterozoic, the Nanhua System (800–685 Ma) and Ediacaran System (formerly the Sinian System, 685–542 Ma) saw the Liaoning landmass uplifted obviously on the northern and southern sides, and the seawater mainly gathered in eastern and southern Liaoning. The strata are called (in ascending order) the Yongning-, Diaoyutai-, Nanfen-, Qiaotou-, Wuxingshan-, and Jinxian Groups. Some stromatolite reefs and metazoan fossils aged ca. 600–570 Ma were found in Dalian-Jinzhou area of southern Liaoning, which is significant for the study of the "Ediacaran-type" biota in Liaoning.

In Cambrian and Middle Ordovician, Liaoning was mainly covered by sea. In the Late Ordovician, affected by the Caledonian movement, the North China Plate, including Liaoning, was strongly uplifted, which lasted nearly 140 million years (460–320 Ma), and caused absence of the Silurian and Devonian deposition. The North China Plate suffered transgression of the Paleo-Asian Sea since the late Early Carboniferous (ca. 320 Ma), and as a result, most areas of Liaoning became coastal or shallow sea, and the Late Carboniferous alternated between marine and non-marine deposits. In Niumaoling of Benxi, eastern Liaoning, there is a typical section showing the Carboniferous (mainly Upper Carboniferous) Benxi Formation, overlying in parallel unconformity the Middle Ordovician Majiagou Formation (limestone), and underlying the non-marine Taiyuan Formation. In the Niumaoling section, there is an obvious unconformity interface in the Middle Ordovician Majiagou limestone, where at the interface, purple iron-bearing shale and G-layer bauxite represent the product of weathering and denudation, which exposed 3–6 m thick and widespread in the whole of North China region. The Benxi Formation was first named by the Chinese geologist Zhao Y. Z. (1925) at the Niumaoling section, and the geologists and paleontologists Li S. G. and Sheng J. Z. also did the research work here.

In the late Early Carboniferous, as Liaoning was located in coastal or shallow water, and climate was warm and humid, lush vegetation gradually appeared in the coastal swamp forest areas, which provided favorable conditions for the formation of coal-bearing deposits. Until the Early Permian, the coal-bearing deposits were distributed in Liaoning, which is called "the first major coal-forming period in Liaoning" and yielded the famous Cathaysian flora.

In the Late Permian, affected by Hercynian movement, the seawater Liaoning had all quit, and sporadic small inland basins formed, mainly distributed in eastern and western Liaoning. In Liaoning, the typical stratum is the lower Zhengjia Formation, which is mainly composed of red sediments (equivalent to the Shiqianfeng Formation). The upper Zhengjia Formation may have entered the Early Triassic of the Mesozoic.

During the transition from Permian to Triassic, the strata in this period were mainly represented by Hongshila Formation or Yangshugou Formation in western Liaoning with red sediments, as in the upper Zhengjia Formation in eastern Liaoning. It was the global drought period, when the vegetation was barren, featured by xerophilous conifers (e.g. *Volt-*

zia), and lycophytes (e.g. *Pleuromeia*). After the Middle Triassic, the non-marine strata and biota were rarely found, except in Linjiawaizi of southwestern Benxi where the Middle Triassic Linjia biota was well preserved and unique in character. The Linjia Formation was mainly composed of gray and yellow-green sandstone and conglomerates with purple and gray-black siltstones and shale, yielding abundant plant and animal fossils. The formation overlay the upper Lower Triassic Zhengjia Formation by parallel unconformity.

About 230 Ma ago, Liaoning entered the Late Triassic, when the climate turned warm and humid. Toward the Early and Middle Jurassic, rivers and lakes were over Liaoning region, and the biota was prosperous, including dense forests. Since the large volcanic activities happened at the time, a large number of sedimentary basins with coal-bearing strata formed. The Early Jurassic coal-bearing strata, such as the Beipiao Formation in western Liaoning, and the Changliangzi Formation in eastern Liaoning, yielded a huge coal deposits, which is called "the second major coal-forming period in Liaoning". The Beipiao Formation was rich in plant fossils, and Mi et al. did a lot of work on the study of Beipiao flora and stratigraphy.

In the Middle Jurassic (ca. 182–163 Ma), volcanic activities, associated with hot and seasonably humid climates, were more intense in Liaoning, and caused many formations of sedimentary basins, spread by lot of rivers and lakes, and all the environments were favorable for lush growth of plants and animals. For example, the Middle Jurassic Tiaojishan Formation in Jianchang of western Liaoning, mainly composed by yellowish grey tuffaceous sandstone and siltstone (ca. 786 m thick), yielded the famous Yanliao Biota. In Tiaojishan Formation of Daxishan in Linglongta, Jianchang, Hu, Xu et al. (2009) first found the earliest feathered dinosaur *Anchiornis*, and Luo, Ji et al. (2012) first found the earliest eutherian mammals *Juramaia*, and a large number of pterosaurs, fish, ostracods, insects and plants were also found. The studies of the Yanliao Biota made important contributions to the research of the origin of birds and eutherian mammals in the world.

After the Late Jurassic relatively arid period (equivalent to the Tuchengzi stage), Liaoning entered the real "greenhouse" time. In the early-middle Early Cretaceous (ca. 135–120 Ma), i.e. the period of Yixian Formation (ca.135–122 Ma) and Jiufotang Formation (ca. 122–120 Ma), Liaoning experienced a strong large-scale volcanic activity again, which formed a lot of sedimentary basins. During the volcanic intervals, in the warm and seasonably humid climates, the rivers and lakes spread all over Liaoning, especially the western Liaoning, where animals and plants were very prosperous, producing the famous Jehol Biota. The biota contained feathered dinosaurs (e.g. *Sinosauropteryx*, *Microraptor*), primitive birds (e.g. *Confuciusornis*), and the oldest known angiosperms (e.g. *Archaefructus*), which provided valuable evidence for the study of the origin and early evolution of many Mesozoic organism taxa in the world.

The evolution of Jehol biota included three stages: the Dabeigou stage (early and cradle stage), the Yixian stage (middle and radiation) and the Jiufotang stage (late and atrophy stage). The Yixian stage is the peak and rapid radiation of the Jehol Biota evolution.

During the late Early Cretaceous (Aptian), it was a warm and humid time in Liaoning. A lot of coal-bearing deposits formed, represented by the Fuxin basin in northern Liaoning, which is called "the third major coal-forming period in Liaoning", dated about 110 Ma.

At the beginning of the Paleogene (ca. 66–50 Ma), a series of rift and uplift occurred in eastern China due to the collision of the Pacific Plate and the Eurasian Plate. The formation of large rift basins in the Lower Liaohe River and Fushun areas constituted graben basins. The volcanic activity at the beginning of the Paleogene was characterized by the basic volcanic rocks of the Laohutai Formation in Fushun, and the volcanic tuffs in the Lizigou Formation. During the intervals of volcanic eruptions, the lakes and marshes in the Fushun basin were in warm and humid climate, forming the coal seam group B in the Laohutai Formation, and the coal seam group A in the Lizigou Formation.

During the Eocene (ca. 50–36 Ma), especially in the middle Eocene, the development of the Fushun biota reached its peak due to the global warming (PETM). The climate of Liaoning was hot, humid, and lush. In the Fushun basin, tropical and subtropical plants, such as *Sabalites* and *Cycad*, were prosperous, forming thick coal seams and oil shale, which is "the fourth major coal-forming period in Liaoning". In particular, the temperature and humidity of Fushun basin were still high during the Jijuntun Formation time (ca. 47.5 Ma). Due to stable subsidence of the basin, the accumulated water deepened into relatively closed lakes, and the hydrodynamic conditions of the lake were weak. In the basin, the lower organisms were prosperous, and their bodies and clay materials formed thick oil shale (up to 48–190 m thick) under the reducing condition by chemical, physical and geological actions. During the late Eocene (ca. 38–36 Ma, represented by the Gengjiajie Formation), the basins were uplifted and the lacustrine basin progressed towards silting due to gradual decrease of temperature. The Fushun biota gradually entered the stage of atrophy.

With the end of the Neogene (23–1.81 Ma) and the beginning of the Quaternary (1.81 Ma–10,000 yrs.), Liaoning basically inherited the early stage of the ancient geography. The Liaohe River area continued to decline, and east and west mountains uplifted. After the end of the ice ages (Pleistocene-Holocene), the Quaternary formation in Liaoning was the accumulation of gravel layer, loess layer, while in some areas volcanic rocks were still seen. Since the Pleistocene, due to new tectonic movement and many other factors, there were moraine, ice water deposits, cave accumulation, and ancient human development. In Miaohoushan of Benxi and the Jinniushan of Yingkou, and some other places the ancient human fossils have been found. The Miaohoushan Man of Benxi (ca. 450,000–500,000 years ago) and the Jinniushan Man of Yingkou (ca. 280,000 years ago) both belong to the stage of *Homo crectus*. Associated with the ancient man fossils, abundant Quaternary mammalian fossils were also found, which belong to the Quaternary *Mammuthus-Coelodonta* Fanna.

1.2 Top Ten Fossil Biotas of Liaoning

Liaoning is one of the provinces yielding most abundant fossils in China, and has the earliest known fossil record of China. So far about 30 categories with over ten thousand species of fossils have been found in Liaoning, including the fossil algae, trilobites, graptolites, archaeocyatha, corals, sponges, bryozoans, brachiopods, cephalopods, conodonts, fusulinids, crinoids, medusoids, gastropods and other marine invertebrates; the amphibians, reptiles, dinosaurs, birds, mammals, fish and other terrestrial vertebrates; the bivalves, gastropods, conchostracans, ostracods, shrimps, insects, spiders and other non-marine invertebrates, and the plants (including sporopollen). The number of fossils ranks first in China.

The abundant fossils mentioned above reveal that during the geological period of nearly 3 billion years, Liaoning experienced the progenitor stage in the Archean Anshan Group, the era of Proterozoic eukaryotes and the multicellular epigenetic (including the Ediacaran stage), as well as the Cambrian-Ordovician biota and the late Paleozoic Benxi biota of the succession of marine biological developmental stages; and also experienced the prosperity and development of the terrestrial biotas from the Mesozoic to the Cenozoic including the Quaternary time. Therefore, the biological evolution in Liaoning appears to be a typical representative of the evolution of life on earth in East Asia.

In view of scientific significance and outputs of Liaoning fossils, Sun et al. (2011) first proposed that the 3 billion years of Liaoning is most characterized by the Top Ten Fossil Biotas, including ①Early Life of Archaen Anshan Group (ca. 2.5–3.0 Ga), ②Cambrian-Ordovician Biota (ca. 460–530 Ma), ③Carboniferous Benxi Biota (ca. 310–330 Ma), ④Middle Triassic Linjia Biota (ca. 240 Ma), ⑤Late Triassic Yangcaogou Biota (ca. 210 Ma), ⑥Jurassic Yanliao Biota (ca. 150–180 Ma), ⑦Early Cretaceous Jehol Biota (ca. 120–140 Ma), ⑧Early Cretaceous Fuxin Biota (ca. 110 Ma), ⑨Paleogene Fushun Biota (ca. 50 Ma) and ⑩Quaternary Ancient Human Biota (ca. 10,000–500,000 years ago).

Among the Top Ten Fossil Biotas, the Mesozoic Jehol Biota and Yanliao Biota, Archean Early Life of Anshan Group and Quaternary Ancient Human Biota are more remarkable and stand out as the "shining points".

1.2.1 Early Life of Archaen Anshan Group

The Anshan Group is one of the oldest known geological records of China, containing the rocks dating back to about 3.8 Ga ago. The fossil cyanobacteria, and iron bacteria, discovered from the metamorphic rocks of the Archean Anshan Group in Xi-Anshan of Anshan and Gongchangling of Liaoyang, have been the oldest known organic record in Liaoning.

In the early life of Anshan Group, the main discovery is of the primitive cyanobacteria fossils which are simple and unicellular in character. In the seawater the cyanobacteria can release oxygen, so that the seawater of ferrous iron (Fe^{2+}) can absorb oxygen into the iron (Fe^{2+}) precipitation (Fe_2O_3), and help form a large number of hematite, and even magnetite.

From the Archaean Anshan Group to the Proterozoic (ca. 2500–542 Ma) the early life gradually evolved into eukaryotes and multicellular epigenetic stages. In Liaoning, the main manifestations are stromatolite fossils found in the southeastern and western Liaoning regions, and which are bio-sedimentary interactive structures constructed by prokaryotic organisms such as cyanobacteria, with the sediments.

1.2.2 Cambrian-Ordovician Biota

After the Nantuo glacial age (ca. 720 Ma), the evolution of early life in China had basically experienced the stages of Lantian Biota, Weng'an Biota, Miaohe Biota, and Gaojiashan Biota, in Precambrian eras. Dating back to 542 Ma ago, it entered the Early Cambrian Explosion time. The Cambrian was mostly prosperous with trilobite, about 60% in number of the entire biota. Since then, it also experienced the Ordovician Biodiversity Radiation event, while the brachiopods, crinoids and tetracorals and other animals grew in the ocean, which ended by ca. 488 Ma ago.

About 540–460 Ma ago in the Cambrian-Ordovician, Liaoning was covered by wide and shallow seas, where the marine biotas were well developed. The Cambrian in Liaoning has not yet been found in the sediments corresponding to the Jinning-, Meishucun- and Nangao formations in eastern Yunnan. The Cambrian fossils in Liaoning are mainly exposed in three regions, including the Taizihe, Lingyuan-Jianchang to the west, and Fuzhouwan, represented mainly by trilobites, associated by archaeocyatha, graptolites, cephalopods, brachiopods, gastropods, conodonts, and sponges, etc. More than 120 genera and 400 species have been found, representing a more stable marine shallow water depositional environments.

Ordovician (ca. 488–443 Ma ago) is the time undergoing extensive transgression. The marine invertebrates were dominant and the biodiversity was greatly enhanced. The Late Ordovician is also the period of volcanic activity, tectonic movement and glacial activity development. The Ordovician in Liaoning is only found in the Lower Ordovician (Yeli- and Liangjiashan formations, mainly of limestone and dolomite) and the Middle Ordovician (Majiagou Formation, of mainly thick-bedded limestone), with absence of the Upper Ordovician strata. The Ordovician fossils in Liaoning are mainly cephalopods, graptolite and conodonts, followed by trilobites, gastropods, brachiopods, etc. The Ordovician conodonts in Liaoning are abundant and belong to the type of the North China, which has been established in fossil sequence zones.

1.2.3 Carboniferous Benxi Biota

The vast areas of the Liaoning and North China plates are basically uplifted from the Late Ordovician to the Early Carboniferous (ca. 458–330 Ma), missing over 100 million years of sedimentary and fossil records. As a result of the decline and uplifting of the North China plate in the Late Carboniferous, Liaoning was once either a shallow sea or the marine and continental alternative in facies during the Late Carboniferous. As the climate was warm and humid, widespread coastal valleys and surrounding terrestrial plants prospered. The ferns, lycopods and seed-ferns began to develop, and the gymnosperms with tall flourishing trees formed the coal-layers in deposits, which is called "the first major coal-forming period in Liaoning", and this coal-forming period was extended to the early Permian, when most of Liaoning had been uplifted to land.

For a long time, Liaoning has been considered to be scarce in the Early Carboniferous deposition. However, since 1990, there are 7 genera and 12 species of plant fossils, including *Sublepidodendron*, found in the lower Benxi Formation by Mi et al. (1990), which evidenced the lower part of the Benxi Formation as the late Early Carboniferous in age.

The Benxi Formation of Carboniferous in Liaoning is about 150 metres thick, with 5–6 layers of marine limestone and several layers of sandstone with coal seams. The typical section of the Benxi Formation is located in Niumaoling, northwestern Benxi. The famous limestone layers in the Benxi Formation include the 5 layers of "lower Mayi limestone", "upper Mayi limestone", "Xiaoyu limestone", "Benxi limestone" and "Niumaoling limestone" in which the conodont fossils studied by Lang et al., indicate the Moscovian in age ca. 310 Ma.

The Benxi Formation was first named by Zhao Y. Z. in 1925, who was the first to find *Spirifer mosquensis* in the Benxi Formation and considered the age of the Benxi Formation as the Middle Carboniferous. Between the 1950s and 1960s, Sheng studied the fusulinid fossils of the Benxi Formation in the Taizihe area, and divided them into 2 biozones and 5 subbiozones, including *Fusulina* and *Pseudostaffella*. Liu (1987) recounted 18 species of 12 genera of brachiopods collected from the lower Benxi Formation, including *Choristites* and *Dictyoclostus*. Lin et al. (1992) systematically studied the coral fossils of the Formation and related 43 species and 3 subspecies of 24 genera, establishing two coral assemblages. During 2005-2007, Sun G., Wang C. Y., et al. carried out a new round of research on the Formation in the Niumaoling section, and got the new findings of conodents including the Moscovian index-fossils *Idiognathodus podolskensis*, *Neognathodus roundyi*, and the assemblage of *Idiognathodus delicates-I. podolskensis*.

In the Early Permian, in Benxi and Nanpiao (western Liaoning) the upper Taiyuan Formation and the Shanxi Formation are mainly continental strata, yielding. abundant plant fossils associated with forming industrial coal seams.

1.2.4 Middle Triassic Linjia Biota

Entering the Mesozoic (ca. 252-66 Ma) Liaoning had been completely turned to terrestrial in facies, and the earth biosphere had great changed to what is called the *Dinosaur Period*. The reptile gave proof to flourishing development, with rapid evolution of other organisms, such as birds, pterosaurs, mammals, and primitive angiosperms, showing an unprecedented prosperity in the biological world in Liaoning.

The unique Middle Triassic biota (ca. 240 Ma) was, found in Linjiawaizi of southwestern Benxi by Zhang et al. (1983), containing more than 30 genera and 40 species of plants, represented by *Symopteris-Benxipteris* assemblage. In recent years, Zhang Y. and Zheng continued the research on the Linjia flora, and newly found *Lobatannularia*, *Pecopteris orientalis*, *Glossopteris*, *Neocalamites*, *Thinnfeldia*, *Glossophyllum*, *Gigantopteris*, etc.

In terms of stratigraphic sequence, the Linjia Formation overlies the uppermost Permian—Lower Triassic Zhengjia Formation characterized by red deposits, and the floristic characters mentioned above, and the age of the Linjia Formation is considered as the Middle Triassic. However, due to the new findings in this formation containing many older plant elements, e.g. *Gigantopteris*, *Lobatannularia*, etc., the age of the Formation may not partly exclude the Early Triassic or late Early Triassic. Thus, the Linjia biota may require further study in age.

1.2.5 Late Triassic Yangcaogou Biota

Yangcaogou Biota was found in the Upper Triassic Yangcaogou Formation (ca. 210 Ma) from the Yangcaogou village of Beipiao in western Liaoning. The flora of this biota consists of more than 36 genera and 71 species, characterized by *Neocalamites*, *Dictyophyllum*, *Ctenis*, *Cycadocarpidum*, etc. The characteristics of the flora indicate a warm and humid climate at that time. The Yangcaogou flora is notable for its large size in specimens, and extremely well-preserved. This flora was studied mainly by Zhou (1981), Zhang & Zheng (1984), etc. According to the study of Zheng (2011*), it can be inferred that the Yangcaogou flora was grown in the subtropical to warm temperate zone of the northern hemisphere and could be roughly classified into the Late Triassic Northern China Floristic Region.

The sporopollen assemblages of the Late Triassic Yangcaogou biota were studied by Qu, Pu and Wu, and as a result, 110 species of 54 genera of palynomorphs were identified. Bivalve fossils have been found in at least 5 genera and 12 species, mainly represented by the assemblage of *Shaanxiconcha longa-Unio ningxiaensis*.

In recent years, Sun et al. (2011) and Sun C. L. et al. (2009, 2014) have studied the Yangcaogou flora again. According to the report of Sun et al. (2011), the flora shows the main representatives include *Annulariopsis?*, *Neocalamites*, *Dictyophyllum*, *Clathropteris*, *Beipiaophyllum*, *Ginkgo*, *Podozamites*, *Cycadocarpidium*, etc., in which the genera *Dictyophyllum*, *Clathropteris* and many of cycadophytes represented the feature of the Late Triassic Southern Flora of China, while the appearance of a large number of ginkgoales (e.g. *Ginkgo*, *Baiera*, etc.) indicated the characteristics of Late Triassic Northern Flora of China. Thus, the Yangcaogou flora seems to show the ecotone features between the Southern and Northern Late Triassic Floras of China.

1.2.6 Jurassic Yanliao Biota

Jurassic is the great development of the Mesozoic biosphere. The Early Jurassic Liaoning, in general, inherited the warm and humid climate of the Late Triassic with lush forest. The Early Jurassic Beipiao Formation represented by coal-bearing strata, indicates "the second large-scale coal formation in Liaoning". However, due to the volcanic activity, geographical changes, and the biological evolution, since the Middle Jurassic, the biota of Liaoning and its neighboring area had undergone major changes, and formed the Yanliao biota. The prosperity of this biota was mainly in the Middle and Late Jurassic (ca. 150–180 Ma) and distributed mainly in the western Liaoning, northern Hebei, and southeastern Inner Mongolia areas.

The name of "Yanliao Biota" was originally derived from the "Yanliao Insect Fauna" named by Hong (1983), with original meaning referring to the insect fauna of Jiulongshan

---

* Zheng S L. 2011. Mesozoic biotas of Liaoning (to be published)

Formation (Hebei) and Haifanggou Formation (Liaoning), aged in the Jurassic. After that Ren (1995) expanded its meaning to cover the Yanliao Fauna, including the Middle Jurassic Haifanggou Formation (Jiulongshan Fm) and Tiaojishan Formation. In 2011, Sun et al. proposed that all the Jurassic biota in western Liaoning and its neighboring areas be referred to as the Jurassic Yanliao Biota. However, after a new round of research, and in this book, Sun et al. suggested that the meaning of "Yanliao Biota" be limited to the Middle-Late Jurassic in age, no longer including the content of the Early Jurassic Beipiao biota.

The Yanliao biota can be roughly divided into two assemblages including (1) Middle Jurassic or Middle-early Late Jurassic assemblage (e.g. in Haifangou, Daohugou, and Tiaojishan Formations) and (2) Late Jurassic assemblage (e.g. in Tuchengzi Formation). More than 20 categories and thousand species of fossils in this biota have been found in recent years. The Yanliao biota in the first assemblage is characterized by the earliest feathered dinosaur *Anchiornis*, and the earliest known Eutherian mammal *Juramaia* in the world. Studies about it have made great contributions to the origin of birds and Eutherian mammals. The pterosaurs and insects were also exceptionally prosperous in the Yanliao biota, and the fossil vegetation changed greatly in composition. All the facts indicate that the climates during the Middle Jurassic were warm and humid, and gradually turned hot and dry since the early Late Jurassic.

The plant fossils of the Yanliao biota (i.e. the Yanliao flora) are generally composed by the three assemblages, including (1) Haifanggou assemblage, (2) Lanqi-Tiaojishan assemblage, and (3) Tuchengzi assemblage. In the Haifanggou assemblage, the mega-fossils are represented by the assemblage of *Coniopteris simplex−Eboracia lobifolia*, in which the genera *Anomozamites*, *Weltrichia* and a large number of conifers *Yanliaoa* are most featured. The palynological fossils are represented by the assemblage of *Cyathidites−Asseretospora−Osmandacidites*, containing 124 species of 48 genera. All the characteristics mentioned above indicate a warmer and humid climate during the early Middle Jurassic time. The second assemblage is characterized by large leaf plants *Ctenis* and *Williamsoniella sinensis*, and gymnospermous pollen *Classopollis* increased more than the first assemblage. In this assemblage, a large number of fossil woods have been found, such as *Lioxylon* and *Araucaria*, *Ashicaulis* and *Protaxodioxylon*. The insect fossils in the Yanliao biota are very abundant, with high diversity. The insects generally are featured by the abundance of Hymenoptera and Diptera (more than 100 species in total).

The Late Jurassic assemblage of the Yanliao biota is represented mainly by those of the Tuchengzi Formation, yielding dinosaurs, fish, conchostraca, ostracods, insects, bivalves, plants, etc. The plants are mainly represented by *Brachyphyllum−Pagiophyllum* assemblage and 39 species, including 5 of wood fossils. Sporopollen fossils are mainly characterized by *Classopollis-Quadraeculina* assemblage, in which *Classopollis* pollen are as high as 57%–82.6% in content.

The dinosaurs in the Tuchengzi Formation are represented by *Chaoyangsaurus yaungi-Jehosauripus*, mainly found in Chaoyang of western Liaoning. In 2015, a large sauropod dinosaur more than 10 m long was discovered in Yabagou village of Beipiao. In the Xiaodonggou Formation of Xinbin in eastern Liaoning, abundant fish fossils (e.g. Ptycholepidae) were also found. The conchostracan fossils in the Tuchengzi Formation are characterized by the assemblage of *Pseudograpta-Beipiaoestheria-Monilestheria*. The fossil ostracods are also rich in the Tuchengzi Formation.

Geologically, the Middle Jurassic stage in Liaoning (e.g. Tiaojishan Formation stage) and the volcanic eruptions were strong and frequent, while in the intermittent period of volcano, the climates were warm and humid, which let the biota flourish in the fluvial and lacustrine environments. However, during the Late Jurassic stage (e.g. Tuchengzi Formation stage), the sediments are mainly a set of gray-green or purple-red interlaced sandstone deposition, yielding a lot of xerophilous plants and other organisms, which meant the climate was warming up and drier than before, and consequently the biota was contracted in developing.

1.2.7 Early Cretaceous Jehol Biota

Jehol Biota is an ancient biota about 1.4–1.2 million years ago living in eastern Asia, including northeast China, Mongolia, Russian Transbaikal, Korean Peninsula, etc. In China, the Beipiao-Lingyuan area and its neighboring region are regarded as the center for its distributions. Previously, the biota was recognized by the assemblage of *Eosestheria-Ephemeropsis-Lycoptera* as the representatives. In the past two decades, due to the discovery of a large number of rare fossils in the Jehol biota of western Liaoning, such as the feathered dinosaur *Sinosauropteryx*, *Microraptor* with four wings, the primitive bird *Confuciuornis*, the earliest known gliding lizard *Xianglong*, the early eutherian mammal *Eomaia* and the earliest known flowers *Archaefructus liaoningensis* and *A. sinensis* and so on, this unique Mesozoic biota has become a focus internationally in paleontology and related sciences. Some important fossil discoveries of Jehol Biota have been the focus of research on the origin and early evolution of birds, eutherian mammals, and angiosperms, etc.

Throughout the past two decades of research on the Jehol Biota, the biodiversity of the Biota in the Early Cretaceous was particularly noticeable. Dinosaurs, birds, pterosaurs, amphibians, turtles, lizards, mammals, fish, insects, spiders, bivalves, gastropods and plants (including wood fossils and sporopollen), etc., more than 20 categories of thousands of species fossils have been found in the Jehol Biota. Only vertebrates have been reported in at least 129 genera and 155 species were found in the Biota, and the fossils were mainly found in the Yixian Formation, and partially found in the Jiufotang Formation.

Dinosaurs

The dinosaurs of this biota, characterized by the small theropod feathered dinosaurs, provide valuable evidence for the research of the origin of birds and their early evolution in the world. Besides them, there are also some larger-sized sauropod dinosaurs found recently in western Liaoning, such as *Liaoningotitan* (MS, ca. 15 m long by 7 m high), and *Dongbeititan*, etc.

Birds

More than 44 species of 38 genera of fossil birds in the Jehol biota have been found in western Liaoning and its adjacent areas. The discoveries demonstrated the first major differentiation and radiation since the emergence of birds in geological times, and revealed some clues on the origin of birds and their flight. The fossil aves in this biota can be divided into three subclasses: Archaeornithes, Enantiornithes and Neornithes, characterized by the genera *Confuciuornis*, *Zhongjianornis*, *Sapeornis*, *Bohaiornis*, *Yanornis*, *Jianchangornis*, *Schizooura* and *Bellulia*.

Pterosaurs

More than 21 genera and 21 species of pterosaurs have been found in western Liaoning, representing an important Early Cretaceous pterosaur group having high disparity.

Other animals

The other important vertebrate fossils in the Jehol biota include the early mammals represented by the earliest known marsupial *Sinodelphys*, and the early eutherian mammals *Eomaia*. Besides, the fossil *Xianglong* represents the earliest gliding lizard in the world. The invertebrate fossils in the Jehol Biota are also very great in number, including nearly 10 categories, such as gastropods, bivalves, conchostracans, ostracods, shrimp, spiders and insects.

Plants

The Jehol Biota is also characterized by the abundant plant fossils, with more than 50 genera and nearly 100 species having been found. The most important finding is the earliest known angiosperm *Archaefructus*. Sun et al. (1998, 2002) confirmed *Archaefructus* as a basal angiosperm, with herbaceous and aquatic in nature. The new discovery and hypothesis the "East Asian centre of angiosperm origin" have put forward the study of angiosperm origin and early evolution in the world.

1.2.8 Early Cretaceous Fuxin Biota

The Fuxin Biota dates back to ca. 110 Ma ago (Aptian) of the late Early Cretaceous. This biota is composed mainly of rich conchastracans, ostrocods, bivalves, fishes, mammals, dinosaurs, and plants, and mainly distributed in Fuxin and Tiefa areas in northern Liaoning.

The flora of this biota is composed of more than 93 species of mega-fossils, represented by the assemblage of *Acanthopteris gothani−Nilssonia sinensis−Ctenis lylata*. The palynological assemblages of *Liaoxisporis−Pilosisporites−Classopollis*, *Pilosisporites−Appedicisporites−Triporolets* and *Deltoidospora−Cicatricosisporites−Appendicisporites* are also very notable.

The animal fossils are characterized by mammals *Mozomus* and *Endotherium*; dinosaurs *Asiatosaurus* and *Heishanoolithus*; fish *Kuyangichthys*, Leptolepidae and *Haizhoulepis*; conchastracans *Yanjiestherites−Pseudestherites* assemblage (still having *Neimengolestheria*); bivalves *Trigonioides−Plicatounio−Nippononaia* assemblage; ostrocods *Cypridea* (*Ulwellia*) *ihsienensis−Limnocypridea qinghemenensis−Protocypretta subglobosa* assemblage (early assemblage), *Cypridea* (*Cypridea*) *tumidiuscula−Pinnocypridea dictyodroma−Mantelliana papulosa* assemblage (middle assemblage), and *Cypridea* (*Pseudocypridina*) *globra− Candona? dongliangensis−Eoparacandona* assemblage (late assemblage); and the insects assemblage of *Hemeroscopus−Cretocercopis*.

1.2.9 Paleogene Fushun Biota

In the Paleogene, after dinosaurs became globally extinct, the appearance of the biological world was completely refreshed. The Paleogene fossils in Liaoning are huge, mainly reflected in the Fushun and Liaohe basins. The biota of this period is collectively referred to as the Paleogene Fushun Biota, aged in the Paleocene-Eocene, about 60−36 Ma ago. This biota includes a large number of plants, fish, insects, conchastracans, ostrocods, turtles and small mammals, with more than ten categories of fossils having been found. The insect fossils are often preserved in the beautiful amber. The Fushun biota is rich in plant fossils, and over 51 genera and 73 species have been found in the Eocene Jijuntun Formation, represented by the assemblage of *Comptonia anderssonii−Cercidiphyllum arcticum−Metasequoia disticha*, including predominant angiosperms (ca. 84%), and gymnosperms (ca. 11%), particularly with the occurrence of *Sabalites chinensis*, *Cycas fushunensis* and some other tropical or subtropical evergreen taxa, which reflect that the climate was quite hot at that time and in comformity with the Paleocene-Eocene global warming event (PETM). Moreover, the lush forests were largely forming coal seams, which is called "the fourth large-scale coal formation in Liaoning". At the same time, in the central and southern Liaoning, the fossil ostrocods, gastropods and algae and other micro-organisms were also very rich, providing favorable conditions for oil forming.

The Fushun biota is also characterized by the insect fossils, mainly preserved in amber and shale of the Guchengzi and Jijuntun formations. The insects are composed of about 201 genera and 223 species belonging to 56 families, represented by the assemblage of *Fushunitendipes eocenicus−Huaxiasciarites longus−Liaoeurytomites petiolatus−Ovalicapito fushunensis−Hypomeces fushunensis*. The other fossils of the Fushun biota, such as conchastracans, ostracods, gastropods, fish, and reptiles are also more or less abundant. The early Eocene conchastracan zone, mainly found in the Xilutian Formation, is featured by *Fushunograpta* assemblage, overlying the Paleocene *Perilimnadia* zone, and underlying the Eocene *Paraleptestheria* zone. The three conchastracan assemblages all appeared to be monotonous, indicating the conchastracans were declining in the Paleogene evolution. The ostracods are mainly found in the Xilutian Formation and occasionally in the Jijuntun and Gengjiajie formations, represented by *Cyprinotus novellus−Cypris bella* assemblage, with about 13 genera and 27 species having been found. It should be mentioned that the fossil turtles characterized by *Anosteira manchuriana* were found only in the oil shale of the Jiuntun Formation.

The micro-group of the Fushun biota, called "Bohai Bay biota" or "Bohai Bay part of Fushun biota" in this book, is also abundant, and mainly found in the Bohai Bay basin near the Liaohe Oilfield. The micro-biota mainly exists in the Paleogene Kongdian Formation (Fangshenbao Fm), Shahejie Formation and Dongying Formation. The fossils are mainly composed of ostracods, gastropods, bivalves, sponges, echinoderms, algae, acritarchs, and sporepollen, and also of fish, mammals, insects, etc. The ostracods have been found to have more than 41 genera and 500 species, including the assemblages of (in ascending order) (1) *Eucypris wutuensis*, (2) *Huabeinia chinensis*, (3) *Camarocypris elliptica*, (4) *Phacocypris huiminensis*, (5) *Chinocythere unicuspidata*, and (6) *Dongyingia inflexicostata*. The Paleogene sporopollen are rich in the biota, mainly found in the Eocene-Oligocene Kongdian-, Shahejie- and Dongying formations. This biota occurs in coastal sea or lacustrine areas, and is affected by sea-flooding, generally with the change of fresh-brackish water environment during the Paleogene.

1.2.10 Ancient human and Quaternary mammal Biota

According to the paleoanthropological research, human evolution can be roughly divided into three stages: *Homo erectus*, early *Homo sapiens*, and late *Homo sapiens*.

The discovery of ancient human fossils in Liaoning began in the 1950s. The earliest discovery was the right humerus bone fossil found in Nandi of Jianping, western Liaoning in 1956, aged in the late late Pleistocene (ca. 15,000 years ago), and belonging to the late *Homo sapiens* in physical characteristics. Since late 1970s, the paleoanthropological study in Liaoning has made great progress, which is accredited to the findings of the earliest known ancient human fossils Miaohoushan Man in Miaohoushan of Benxi, eastern Liaoning, aged in the middle Pleistocene (ca. 450,000−500,000 years ago), and the "Jinniushan Man" in Jinniushan of Yingkou, southern Liaoning, aged in the middle Pleistocene (ca. 280,000 years ago), and other stages of the ancient human fossils, with more than 20 Paleolithic cultural relics. These discoveries represent three stages of the Paleolithic Age. So far in Liaoning, more than 80 fossil sites of the ancient human and Quaternary mammals have been found, including also the Gezidong Man (Shuiquan of Kazuo, western Liaoning, the middle-late late Pleistocene, ca. 50,000−70,000 years ago), Xiaogushan Man (Gushan of Haicheng, southern Liaoning, the late late Pleistocene, ca. 40,000 years ago), and Qianyang Man (Qianyang of Dandong, southeastern Liaoning, the late late Pleistocene, ca. 18,600 years ago), etc.

The Quaternary mammals basically belong to *Mammuthus-Coelodonta* fossil fauna, distributed in more than 80 fossil sites in Liaoning. The ages of the Quaternary mammals include the early Pleistocene (ca. 3−1 Ma ago), the middle Pleistocene (ca. 1 Ma−200,000 years ago), and the late Pleistocene (ca. 200,000−10,000 years ago). The typical Quaternary mammal faunas include the Haimao fauna (Dalian, 1.6−1.2 Ma), Miaohoushan fauna (Benxi), Jinniushan fauna (Yingkou), Maoershan fauna (Kazuo), Anping fauna (Liaoyang), Zangshan fauna (Yingkou), Gezidong fauna (Kazuo), and Mashandong fauna (Chaoyang), and so on.

Chapter 2

Introduction to Paleontological Museum of Liaoning

Liaoning is one of the "fossil treasures" of China, and the world at large. The study of fossils of Jehol biota and Yanliao biota of Liaoning has made great contributions to the research of the origin and evolution of life in the world. Thus Liaoning is called "the place where first bird flew and the first flower bloomed". Fossils of the early life in Anshan biota trace the geological history of Liaoning to about 2.5–3.0 Ga ago, which indicates Liaoning Province has the oldest known fossil record in China.

In order to further protect and study the unique fossil resources in Liaoning Province, the provincial government of Liaoning approved of the plan to co-construct the Paleontological Museum of Liaoning (PMOL) by Department of Land and Resources of Liaoning Province (DLRL) and Shenyang Normal University (SNU) in 2006. After five years effort in construction, the largest known paleontological museum in China was officially opened to the public on May 21, 2011.

PMOL is located near the main gate of the SNU, with a building area of 15,000 m^2. The building looks like a huge geological body and a giant dinosaur in shape, with a fault cutting the geological body vertically, and the volcanic lava flowing and pouring from top to bottom, taking us back to the geological history in Liaoning. The southern part of the building represents the giant dinosaur with its egg, and the 21 towering steel frame which works like the dinosaur ribs and symbolizes a bright future for Liaoning in the 21st century.

The PMOL focuses on science, and showing the origin and evolution of life history is the main line in the exhibition. The museum is characterized by the Top Ten Fossil Biota in Liaoning through three billion years, and by its distinctively international nature. The museum consists of 8 halls and 16 exhibit areas, prominently displaying the Early Life of Anshan Group, Jehol Biota, Yanliao Biota, and Paleoanthropology in Liaoning. The Museum pays great attention to both scientific research and science popularization, upholding the idea that science should serve the public. Through the five years effort, the PMOL has already become one of the national paleontological centers for scientific research, science popularization and paleontological teaching in China, particularly for Liaoning. The Hall 4 introduces the world-wide famous Jehol Biota, bringing the audience into the history hundreds of millions of years ago in western Liaoning, with the bright exhibitions of the Dinosaur Kingdom, Primitive Avian World, Cradle of Flowers and Associated Biological Groups. The Hall 8 is a magnificent hall of Giant Dinosaurs in Liaoning, in which there are eight huge dinosaurs notably marked by the giant dinosaur *Liaoningtitan*, about 15 m long, found and studied by the PMOL for the first time.

The establishment of the PMOL is a major event in the cultural and educational undertakings of Liaoning Province and China's paleontology. It is of great significance to enhance the level of scientific research and popularization, and strengthen the fossil protection in Liaoning. Meanwhile the PMOL promotes the international exchange and cultural tourism in Liaoning, and carries forward the study of paleontology in Northeast China. In September 2012, when Mr. Xu S. S., Ex-Minister of Ministry of Land and Resources, China (MLRC), visited the museum, he admired the establishment of the PMOL as a successful example for the cooperation between the provincial government and university to co-build museum in China.

2.1 The PMOL design and exhibition

The structure of the PMOL is like a fine work of art: a clever fusion of a huge geological body with a giant dinosaur-like steel frame, showing a upright dinosaur borne from its mother, the Mesozoic strata. In the middle of the separation from the fault, volcanic lava pours down along the fault on both sides, bringing us back to the land of Liaoning more than 100 million years ago, when the organisms competed with the earth, and were interdependent in the geological periods. The architectural design of the PMOL is like a giant art sculpture, leaving magnificent and spectacular impression on the audiences.

The building exterior is designed by the famous architect, Mr. Li Z. Y. from the Architectural Design Firm of Taiwan, China. Mr. Li is a well-known architect, a graduate from Taiwan Cheng Kung University who received Master's degree in architecture in Princeton University, US. He has been dedicated to researching and creating new buildings that inherit traditional Chinese features, providing life-building as a design concept, combining modern technology to provide high-creative, highly integrated and high-tech architectural design services. He took charge of the design of Taipei 101 Building, which was once the third largest skyscraper in the world. The exterior design of PMOL building can be regarded as one of his peerless masterpieces.

When you enter the PMOL, you will see the entire exhibition run through the Evolution of Life as the main line, highlighting in the four major parts, including the Jehol Biota, Yanliao Biota, Early Life in Anshan Group, and Ancient Humans in Liaoning. The layout of the exhibition halls is as follows:

Hall 1 (Zone Ⅰ)—Introduction
Hall 2—Earth and Early Life
  Zone Ⅱ: Origin of Earth & Life
  Zone Ⅲ: Cambrian Explosion
Hall 3—Fossils of Liaoning through 3 billion years
  Zone Ⅳ: Outline of Geology of Liaoning
  Zone Ⅴ: Top Ten Fossil Biotas of Liaoning
Hall 4—The Jehol Biota
  Zone Ⅵ: Introduction
  Zone Ⅶ: Dinosaur Kingdom
  Zone Ⅷ: Primitive Avian World
  Zone Ⅸ: Cradle of Flowers
  Zone Ⅹ: Associated Fossils
Hall 5—International Paleontology in the World
  Zone Ⅺ: History of Paleontology
  Zone Ⅻ: Fossils in the World
  Zone ⅩⅢ: Extinction of Dinosaurs and the K-Pg Boundary
  Zone ⅩⅣ: International Cooperation & Exchange in Liaoning
Hall 6—Interaction in Science for Popularization
Hall 7 (Zone ⅩⅤ)—Rare Fossils
Hall 8 (Zone ⅩⅥ)—Giant Dinosaurs of Liaoning

In Hall 1, Brief Introduction of PMOL, the geological gallery presents examples of the history of geological evolution in Liaoning through 3 billion years. From the colorful rocks of geological section, you can get a general picture of the biostratigraphical changes of northern China during the geologic history, including the Archean, Proterozoic, Paleozoic, Mesozoic and Cenozoic periods. The geological profile corresponds to the paleoenvironmental background, leaving a preliminary impression on audiences, particularly in the Yanliao biota, Jehol biota, Ancient human in Liaoning, and the four periods of large-scaled coal-forming in Liaoning.

Enter Hall 2, Earth and Early Life, to experience the slow and difficult emergence and evolution of early life through 4.6 Ga, including the introduction to the evolution of the original oceans from the inorganic to the organic phase (ca. 4.6–3.8 Ga ago), the emergence and early evolution of the Metazoan, such as the Edicaran biota, Cambrian Explosion in Chengjiang biota (China), and so on.

Hall 3 Fossil of Liaoning throught 3 Ga, and Hall 4 Jehol Biota are the two special halls in the PMOL. Hall 3 focuses on the bio- and geological history of Liaoning, featured by the "Top Ten Fossil Biotas in Liaoning", including the Early life of Archean Anshan Group, Cambrian-Ordovician marine biota, Carboniferous Benxi biota, in the Paleozoic and before; the Middle Triassic Linjia biota, Late Triassic Yangcaogou biota, Yanliao biota, Jehol biota and Fuxin biota, in the Mesozoic; and the Fushun biota, and the Ancient human and Quaternary mammal biota in Liaoning, in the Cenozoic. Hall 4 the Jehol biota, has a great variety of amazing and beautiful specimens to shine brilliantly in the four exhibitive zones, including the Dinosaur kingdom, Primitive avian world, Cradle of flowers, and Associated fossils.

Hall 5 International Paleontology in the World presents the fossils mainly from Germany, UK, US, Russia, Japan, Australia, India, Thailand, Afghanistan, Estonia, and other countries. Of the fossils, 24 precious fossils (casts) from the famous Messel Eocence biota were donated by the Senckenberg Natural History Museum of Germany. In Hall 6 Interaction in Science for Popularization, visitors can interact with museum by Racing with Dinosaur, Photographing with Dinosaur, Dropping Dinosaur Eggs, Staying in Dinosaur Theater, watching movies in the 3D-Cinema, etc.

Hall 7 Rare Fossils displays the treasures of the PMOL fossils, including *Anchiornis huxleyi*, *Megaconus*, *Xianglong*, *Shenshiornis*, *Dapingfangornis*, *Liaocheys*, *Archaefructus* and *Psittacosaurus*, which show the amazing fossil of 39 juvenile individuals called *Psittacosaurus* Kindergarten. All the rare fossils feast the audiences' eyes and win their praise for scientific significance. Hall 8 Giant Dinosaurs of Liaoning shows eight large dinosaurs collected in western Liaoning for the first time, represented by *Liaoningtian* (MS) which attracts a constant swirl of visitors who are reluctant to leave.

The inner exhibition design of the PMOL is completed personally by experts of the museum led by Prof. Dr. Sun G., well-known paleontologist in China and Director of the PMOL. The ideas on the exhibition mainly include scientificity, internationalization, and emphasizing Liaoning. The design also fully takes into consideration the exhibition for popularization and the propaganda on fossil protection in Liaoning, and the application of high-tech means, as well as the participation and advices of the scientists.

For the scientificity, the design emphasizes the biological evolution as the main line. It promotes Darwin's theory of evolution of life in the exhibitions, conveys the geological and paleontological knowledge in a high level, and pays attention to the fossil and related display with scientific connotation and academic support. In this exhibition, combined with Liaoning fossils, the guiding principles are to highlight the origin and early evolution of life, such as the significance of feathered dinosaurs in the study of bird origin, of the earliest known flower fossils in the study of angiosperm origin and early evolution, and of the discovery of ancient human fossils in Liaoning for the study of human evolution and migration in Northeast Asian region. The exhibition, supplemented by scientific works (e.g. published in *Science* and *Nature*), examples of scientific experiments and the latest reconstructions of the paleobioenvironments, such as the Yanliao biota and Jehol biota, enables the audience to have a more profound understanding.

For the internationalization, PMOL has been learning from the world's leading natural history museums (e.g. the Natural History Museum of London, Smithsonian Museum of Natural History, American Natural History Museum in New York, Senckenberg Natural History Museum, and Fukui Prefectural Dinosaur Museum of Japan, etc.), and as a result, the international advanced design concepts have been integrated into the exhibition of PMOL. At the same time, in order to strengthen international exchange and cooperation and to stimulate the interest of audience, the exhibition increases the number of fossil specimens and related objects from the world, and with the introduction of international scientists, and their achievements, PMOL can surely widen the audience's international perspective in paleontology and geosciences.

The design idea for emphasis on Liaoning mainly stresses the introduction of the Top Ten Fossil Biotas in Liaoning in the past 3 billion years, the characteristics of the Jehol biota, Yanliao biota, early life, and ancient human in Liaoning, and highlighting the "four stars" of valuable fossils in Liaoning, such as *Anchiornis huxleyi, Xianglong zhaoi, Archaefructus* and *Shenshiornis*. Meanwhile, the design has taken into consideration that nearly whole exhibition specimens were collected from Liaoning, and they can be especially shown in Hall 8 Giant Dinosaurs of Liaoning to display the eight large dinosaurs, markable by the biggest known sauropod dinosaur in Liaoning, *Liaoningtitan* (MS), found and studied by the PMOL experts in Liaoning.

On February 21, 2009, the Evaluation Meeting on Outline of Design for Exhibition of PMOL was held in Shenyang, co-sponsored by DLRL and SNU and joined by the famous paleontologists and experts in museology in China. The meeting was chaired by Academician Prof. Liu J. Q., and the experts including professors Dong Z. M., Xu X. and Jin C. Z. (IVPP), Ji Q. and Lv J. C. (CAGS), Gao K. Q. (PU), Wang W. M. (NIGPAS), Sun C. L.,

Duan J. Y. and Xu Y. (JU), Zheng S. L., Wang W. L. and Zhang L. J. (SIGM), Peng G. Z. (Zigong DM), Cui B. (GMH), etc. All the experts highly appraised this design outline, and admired the exhibition design as reaching advanced level internationally.

2.2 Review of PMOL in the past five years

The successful co-construction of the PMOL is an example by Liaoning provincial government (DLRL) with university (SNU) in implementing the National Regulation on Fossil Management, as well as a landmark created by their joint efforts. The museum construction not only meets the need of fossil protection and the study of China but also wins the confidence of scientists and people both at home and abroad.

2.2.1 Intense connection

In October, 1996, the Chinese journal first published the paper written by Ji & Ji, reporting the earliest discovery of feathered dinosaur *Sinosauropteryx* in China, which caused a sensation in international academic circles. In November, 1998, the American journal *Science* published the paper by Sun et al. on the first discovery of the earliest known angiosperm *Archaefructus* in western Liaoning, China, which further drew worldwide attention to Liaoning. With the continuous discoveries of *Confuciusornis*, *Caudipteryx*, *Microraptor* and other fossils found from western Liaoning, which show amazing academic value, particularly their vital role in the study of organic evolution in China, and even the whole world, and precious fossils of Liaoning have drawn more and more attention from international academic circles. Therefore, Liaoning, honored as "the place where the first bird flew and the earliest flower blossomed", becomes well-known all over the world.

In consideration of discoveries of so many important fossils, the fossil protection was put on the agenda of the DLRL officially. In 2001, the DLRL issued a local regulation—Regulation on Fossil Protection in Liaoning Province which was approved by the Liaoning Provincial People's Congress. To better protect and study fossils in Liaoning, Ex-Director of the DLRL, Mr Wang D. C., proposed to choose SNU as the best collaborator, which was fully agreed and supported by the SNU, particularly by Prof. Zhao D. Y., Ex-President of the SNU. Thus, both sides reached a consensus immediately.

Shenyang Normal University (SNU) is a comprehensive university with a history of over 50 years. It offers many disciplines, i.e. philosophy, economics, law, education, liberal arts, science, engineering, management, arts and many others, and plays a vital role in education, scientific research and social service within the province. SNU has always prioritised fossil protection in Liaoning and the need for scientific research. It is fortunate that SNU has a lot of experts in biology, and the Institute of Mesozoic Fossil Research of

Western Liaoning is guided by Prof. Hou L. H., the famous avian paleontologist from IVPP, and since 2007, Prof. Sun G., the famous palaeontologist, has been Director of the Institute plus joint work of a batch of young paleontologists with Ph.D. degree from the institutions of Japan, Australia, Peking University (PU) and so on. Thus, SNU was capable of taking responsibility, and promised to build the new museum with the DLRL. Therefore, in Liaoning, the PMOL construction was jointly started by government and university, with a ceremony in the SNU campus on June 6th, 2006.

2.2.2 Create the great cause together

With the joint efforts by SNU and DLRL, after five-year hard work, a mansion, the PMOL about 30 m high and 15,000 m^2 in building area, stood erect on the ground. It was an epoch-making event for Liaoning to have its first palaeontological museum.

During the five years, leaders of the two construction sides, SNU and DLRL, made enormous efforts in the PMOL construction. Ex-President of SNU Prof. Zhao D. Y. and the director of DLRL Mrs. Ji F. L., and other leaders very often came to the construction sites to give instructions to the construction work, and take care of everything about the PMOL in construction. Meanwhile, the PMOL construction won support and care from the leaders of provincial government, relevant national ministries and many experts, such as Lu X., Chen X. and Chen C. Y. (Ex-Vice-Governors of Liaoning), Jiang C. S. and Shou J. H. (Ex-Vice Ministers of MLRC), and Chinese academicians Li T. D., Liu J. Q., etc.

During the five years of construction of the PMOL, sincere care and efforts have also been given by international palaeontologists and museum leaders, represented by Dilcher D. L., (NAS, Ex-President of BAS, US), Edwards D. (President of Linnaean Society, UK), Akhmetiev M. (Corr. Mem. RAS, Russia), Mosbrugger V. (Academician, and General Director of Senckenberg Institutions, Germany), Eder J. (President of IOP and Director of Stuttgart Museum of Natural History, Germany), Johnson K. (Director of Smithsonian National History Museum, US), who all came to construction site of the museum, and not only praised the great cause of construction but also gave valuable suggestions. In addition, the German Senckenberg Museum donated 24 precious replicas of fossils of the Messel to the PMOL.

In a word, the successful construction of the PMOL is not only the fruit of wisdom and hard work of so many designers, constructors, managers and scientists, but also a proof showing that China, in construction of natural museums, won support and understanding from international academic circles.

2.2.3 Opening ceremony

On May 21, 2011, the opening ceremony of the PMOL was held in the SNU campus and the museum. The government leaders, Vice-Ministers Lu X. (MOEC) and Wang M. (MLRC), Vice-Governor Chen C. Y. (Liaoning), and experts academicians Li T. D., Liu J. Q. and Zhou Z. H., American NAS Dilcher D. L., and many scientists from the US, the UK, Germany, France, Russia, Japan, Mongolia, Korea, Vietnam, India, and Brazil, along with Chinese experts, attended the ceremony and witnessed the opening of the PMOL, and met audiences from both at home and abroad.

2.3 Ten Fossil Stars

At the time when the PMOL opened, the four precious taxa including *Anchiornis, Shenshiornis, Xianglong* and *Archaefructus*, were named as the Four Stars in exhibition of the museum. In recent years, with the increase of fossil collection and study, more precious fossils were continuously found in Liaoning. Therefore, this book proposed a notion of the Ten Fossil Stars to supplement the former Four Stars. They are as follows: *Anchiornis, Liaoningotitan* (MS), *Psittacosaurus* ("kindgarden"), *Megaconus, Xianglong, Archaefructus, Confuciusornis, Dapingfangornis, Liaochelys*, and *Shenshiornis*.

(1) *Anchiornis* Xu, 2009: The earliest feathered dinosaur in the world

Anchiornis was initially named by Xu et al. in early 2009. Hu D. Y. et al. reported the second specimen of the genus in *Nature* on Oct. 1, 2009, which first confirmed this specimen representing the earliest known feathered dinosaur in the world.

Anchiornis has nearly complete feather preservation. Long pennaceous flight feathers were attached to the forelimbs and hind limbs, forming the earliest known "four-wing" dinosaur. Shorter feathers were even attached to the pedal phalanges apart from the unguals. *Anchiornis* was found in the Middle-Late Jurassic Tiaojishan Formation of Jianchang, Liaoning, dating back to ca. 160 Ma, which is earlier than the German *Archaeopteryx* by over 10 Ma, representing the earliest feathered species in the world. Hu et al. proposed *Anchiornis* belongs to a theropod group known as the troodontids, which powerfully evidences the hypothesis that birds originate from theropod dinosaurs, and also supports the hypothesis that the evolution of bird flight experienced a "four-winged" stage. Hu et al. also inferred that all major theropod groups, including Aves, might have originated and diversified rapidly in the Middle to the earliest Late Jurassic. This contribution was selected as Top Ten Achievements of Science and Technology in Chinese Universities, 2009 and Top Ten Scientific News in China/World, 2009, and the new finding is honored as "bridging a critical gap between dinosaurs and birds" by international academic circle.

Locality & Horizon: Daxishan of Jianchang, Liaoning; Tiaojishan Formation (ca. 160 Ma, Middle-Upper Jurassic)

(2) *Liaoningotitan* (MS): The biggest dinosaur in Liaoning

Liaoningotitan is a new taxon of Titanosauriformes and the largest known dinosaur found in the Lower Cretaceous of Liaoning. It is nearly completely preserved. This dinosaur is about 15 m long; its forelimb is relatively shorter, as about 70% of the hindlimb. Skull and lower jaws are well preserved. The buccal margin is convex; the jugal is anteriorly positioned, close to the anterior edge of antorbital fenestra. Maxillary teeth are rostrally positioned and imbricate. The crown is narrow with a D-shaped section. Nine lower jaw teeth are relatively small, and well spaced. The lower tooth crown is asymmetrical with an oval section; the lingual groove and ridge are developed, while the lingual bubble-like eminences are developed at the base of crown. As other sauropods, *Liaoningotitan* also has a long neck and tail, and a large body. The fossil was unearthed and studied by the PMOL staff since 2006. This discovery not only enriches the diversity of dinosaurs in the Jehol Biota, but also provides new evidence in understanding the early evolution and radiation of Titanosauriformes.

Locality & Horizon: Beipiao of Liaoning; Yixian Formation (Lower Cretaceous, ca. 125 Ma).

(3) *Psittacosaurus*: A juvenile nest as "dinosaur kindergarten"

Psittacosaurus is a herbivorous ceratopsian dinosaur with gregarious and parental care behavior and found in East Asian region. *Psittacosaurs* are generally less than 2 m long. The skull is rectangular in lateral view, and triangular in dorsal view. The snout is high; the jugal horns are laterally extended. The forelimb is relatively short and hindlimb is longer and stronger. The Psittacosaur nest shows 39 juveniles, which is the largest nest of Psittacosaur known in the world, named as "dinosaur kindergarten". These juveniles varied from 25 cm to 40 cm in length, and gathered in 0.8 cm^2 in area, which is significant for the study of psittacosaurs in taphonomy and ecology. Moreover, the museum also exhibits a sample called "mother with child" which shows the parental care behavior.

Locality & Horizon: Beipiao of Liaoning; Yixian Formation (Lower Cretaceous, ca. 125 Ma).

(4) *Megaconus* Zhou et al., 2013: A primitive haired mammaliaformes

Megaconus is a primitive mammaliaformes, with the type species *M. mammaliaformis*, about 30 cm in length, and 250 g in weight. Its upper molars have longitudinal cusp rows that occlude alternately with those of the lower molars. This specialization for masticating plants indicates that herbivory evolved among mammaliaformes, before the rise of crown mammals. Its mandibular middle ear and primitive ankle are the ancestor characters of mammaliaformes. But the molars of *Megaconus* are highly specialized with healed high-crowned root and precisely occluded teeth, which shows that mammaliformes had advanced tooth types and functional adaptation of feeding habits. The ends of the tibia and fibula are healed, as in the extant armadillos or rock hyrax. The preservation of skull and skeleton, even hairs in *Megaconus* is only the remains for the most primitive Haramiyida, and very important discovery for study of the origin of mammals.

The research result was published on *Nature* in 2013 by Prof. Zhou C. F. and his research group, including Prof. Luo Z. X. (US) and Prof. Martin T. (Germany).

Locality & Horizon: Daohugou of Ningcheng, Inner Mongolia; Daohugou Formation (Middle Jurassic, ca. 165 Ma).

(5) *Xianglong* Li et al., 2007: The oldest gliding lizard in the world

Xianglong, with the type species *X. zhaoi*, is the newly found lizard fossil in the Jehol biota. In evolution history of lizard over 200 million years, *Xianglong* is the only gliding lizard fossil. It is 15 cm long, with 8 prolonged ribs laterally to support patagium. *Xianglong* could climb in the forest and also glide in the sky, as the extant flying lizard in southern China and Southeast Asia. This discovery fills the blank of gliding in evolution history of lizard in the Early Cretaceous, as well as provides better evidence for studying tropical or subtropical paleoenvironment of the Jehol biota.

Prof. Li P. P. led this research, and the paper was published on *PNAS* in 2007.

Locality & Horizon: Beipiao of western Liaoning; Yixian Formation (Lower Cretaceous, ca. 125 Ma).

(6) *Archaefructus* Sun et al., 1998: The earliest known flower in the world

Archaefructus is an aquatic herbaceous angiosperm, and the earliest known angiosperm in the world, including *A. liaoningensis* and *A. sinensis*. *Archaefructus* has simple determinate axes bearing helically conduplicate carpels enclosing several ovules in each; stigmas not differentiated, and stamens often paired with monosulcate pollen. Leaves are thin, highly dissected with delicate stems, root system simple and poor in development, and absent of perianth. All the characteristics indicate *Archaefructus* is a primitive angiosperm and aquatic and herbaceous in nature, which is significant for the study of angiosperm origin maybe as aquatic. In 1998, Sun G. proposed the hypothesis of the "East Asian center of angiosperm origin". The papers on *Archaefructus* were published in *Science* as cover paper for two times, and selected into Top Ten News of Science & Technology in China for 1998, and Top 100 Science News in 2002 (*Discover*, US), which put forward the study of origin and early evolution of angiosperms in the world.

The study has been led by Sun G., Dilcher D. L. and their research group.

Locality & Horizon: Beipiao and Lingyuan of Liaoning; Yixian Formation (Lower Cretaceous, ca. 125 Ma).

(7) *Confuciusornis* Hou et al, 1995: A primitive bird found in west Liaoning

Confuciusornis is one of the primitive birds earliest discovered in Liaoning. It was named by Hou et al. in 1995. *Confuciusornis* has a crow-sized body, a toothless horny beak as in modern birds, and a short tail in which the final caudal vertebrae are fused to form a pygostyle, but it still retains a typically diapsid skull and large manual unguals on the forelimbs like small theropod dinosaurs. The discovery of *Confuciusornis* filled the blank of the study on the early evolution of birds in the world after *Archaeopteryx*.

Locality & Horizon: Western Liaoning and its neighboring areas; Lower Cretaceous Huajiying Formation (131 Ma), Yixian Formation (ca. 125 Ma) and Jiufotang Formation (ca. 120 Ma).

(8) *Dapingfangornis* Li et al., 2005

Dapingfangornis, with type-species *D. sentisorhinus*, is a small-sized enantiornithine bird, unearthed by the PMOL at Dapingfang of Chaoyang, western Liaoning in 2005, and studied by Li L. et al. in 2006.

The specimen is a complete skeleton, with a body of about 22 cm long and a skull of 2.5 cm long, having long feather crown on the skull and two long tail feathers, and a high bone crest on the nasal bone. Both upper and lower jaws have many teeth. The forelimbs still retain manual unguals. Its long and curved pedal unguals and fully reversed hallex imply a typical arboreal life type. This discovery provides important evidence for the evolution of bony and feathery ornamentation of early birds.

Locality & Horizon: Dapingfang of Chaoyang, Liaoning; Jiufotang Formation (Lower Cretaceous, ca. 120 Ma).

(9) *Liaochelys* Zhou, 2010

Liaochelys is the third turtle taxon found in the Jehol biota, and bears a close phylogenetic relationship with *Manchurochelys* and *Ordosemys*. The genus is abundant in the Jehol biota, and belongs to the family Sinemydidae, featuring low carapace and plastron with a ligamentary connection.

Liaochelys is characterized by the nearly elliptical carapace, short and wide vertebral scales, two suprapygal plates in which the first plate is obviously smaller than the second, and the distally expanded third costals.

Liaochelys was studied and named by Prof. Zhou C. F.

Locality & Horizon: Jianchang of Liaoning; Jiufotang Formation (Lower Cretaceous, ca. 120 Ma).

(10) *Shenshiornis* Hu et al., 2007

Shenshiornis belongs to Sapeornithids, a unique group only found in the Early Cretaceous Jehol biota. Sapeornithids have the largest body size, presently known among the Early Cretaceous birds, with a pair of super long forelimbs.

Shenshiornis prima, as the type species of the genus, is 40 cm long, having a typically diapsid skull with a high and stout snout as dinosaur, and the upper jawtip still retains several stout conical teeth, showing a streamlined and kinetic skull evolved later in avian evolution. However, its hind limbs were highly modified for arboreal life; the tibiotarsus became relatively short, and the hallex was fully reversed backwards. The discovery of *Shenshiornis* provides important evidence for the early revolution of kinetic skull of birds and arboreal life.

Shenshiornis was discovered by the PMOL in October, 2005, and studied by Hu D. Y. et al. during 2006–2010. This generic name *Shenshi-* is given to honor Shenyang Normal University which is referred to as "Shen-Shi" in Chinese lauguage for short.

Locality & Horizon: Dapingfang of Chaoyang, Liaoning; Jiufotang Formation (Lower Cretaceous, ca. 120 Ma).

2.4 Fruitful scientific research

The PMOL takes scientific research as an important criterion for testing the scientific level of the museum. In view of its own advantages of the PMOL, the museum has paid more attention to the research areas, including the paleobotany (mainly in early angiosperms), avian paleontology, fossil amphibians and mammals, and paleoanthropology in Liaoning, which pointed the directions in research and won a batch of important results.

2.4.1 Major achievements in the research

In recent years, researchers of the PMOL published over 100 papers, of which more than 40 were indexed by SCI; 6 papers were published on *Nature* (UK, including 3 as the first author); 2 papers appeared on *PNAS* (US) and others on the important journals, e.g. *JVP, IJPS, BMC, Science China*. Moreover, the museum had 4 monographs and 2 textbooks published, such as *The Fossil Record of Three Billion Years in Liaoning* (2011), *Late Cretaceous-Paleocene Biota and the K-Pg Boundary from Jiayin, Heilongjiang, China with Discussion on the Extinction of Dinosaurs* (2014); *Early Paleozoic Faunas, Facies and Mul-*

tiple Stratigraphic Divisions in Eastern North China Plate (2015). For the past five years, the museum has received 36 research projects given by the NSFC and other financial resources from ministerial level. The achievement "Study of early angiosperms in Laioning" won the Scientific & Technical Award of Liaoning Province (in first grade), and the "Discovery of the earliest feathered dinosaur *Anchiornis*" was selected into Top Ten Advances in Science & Technology of Chinese Universities in 2009 and Top Ten Scientific News in China/World in 2009 in a ballot by Chinese academicians. All the achievements have greatly helped the study of origin and early evolution of organisms in China and the world.

2.4.2 Other achievements in research

The main other achivements include also the discovery of a batch of fossil enantiornithean birds in Liaoning, represented by *Xiangniaornis* (Hu et al., 2013) and *Shengjingornis* (Li et al., 2012), published in *JVP* and *Act. Geol. Sin.*, respectively; the discovery of new taxa of pterosaur *Jianchangnathus* (skull) in Jurassic Liaoning, and turtle *Xiaochelys* in Liaoning's neighboring area, published in *JVP* and *Sci. Rep.*, respectively (Zhou et al., 2014, 2015); the first definition of the non-marine K-Pg boundary in Jiayin of Heilongjiang, China, published in *Global Geology* (Sun et al., 2011) and the monograph (Sun et al., 2014); and the discovery of Osmundaceae mineralized rhizome fossils with two new species of *Ashicaulis beipiaoensis* and *A. wangi*, published in *IJPS* and *Sci. China* (Tian et al., 2013, 2014).

2.5 Active scientific popularization

"Science should serve the public" is the idea held by the PMOL, and carried out in the scientific popularization activities. Since May, 2011, nearly a million people have visited the PMOL, with up to 7,000 persons/day in vacations, and nearly 4,000 social groups from various circles. The museum cooperated with 71 middle and primary schools for scientific popularization, and engaged more than 800 student volunteers as assistants for the museum. The popularization programs include "Little interpreter training", "Wonderful night in museum", "Youth class", "Exhibition competition of child science show" and so on. Also, it developed programs such as "Making friends with dinosaurs", "Collecting fossils", "Protecting environment, cherishing resources", etc. Moreover, the museum holds some memorial activities with distinct themes on important memorial days such as World Earth Day, Environment Day, and Museum Day.

To promote scientific popularization, the museum had some books published, such as *The Fossil Record of Three Billion Years in Liaoning* (Sun et al., 2011), *Entering the*

Hometown of Birds (Sun et al., 2013), and *Life Evolving from Prehistory Time* (Yang et al., 2015). The first and the third book won the Award for Excellent Books for Scientific Popularization given by the MLRC in 2014 and 2016, respectively.

For five years, the PMOL has successively been granted more than ten honors in scientific popularization bases, such as the National Popular Science Base of China Association for Science and Technology, National Popular Science Base of Palaeontological Society of China, National Popular Science Base of MLRC, Liaoning Provincial Popular Science Base, and Shenyang Scientific Education Base for Youth. Moreover, the museum was awarded the National Model Natural Museum in China in 2015.

2.5.1 Special exhibitions

Holding the special exhibitions is one of distinctive scientific popularization activities by the PMOL. To better carry out scientific popularization and to fully exploit the advantages of fossil resources in Liaoning and the experts of the PMOL, since 2011, the museum has held four major special exhibitions which were on Liaoning Dinosaurs (2012), Fossil Plants in Liaoning (2013), Special Exhibition in the Horse Year (2014) and From Ape to Man (2016). Besides, the museum conducted several special exhibitions on Feathered Dinosaurs in Liaoning, China abroad.

2.5.2 Experts' participation and guidance

Expert' participation with communication to visitors is one of features of the PMOL and this gives the exhibition more profound cultural connotation. For instance, the Special Exhibition of Liaoning Dinosaurs (2012) was attended by several famous experts such as Xu X. and Hou L. H. (IVPP), and Ji S. A. (CAGS), which brought a big surprise and stimulated the interest of the visitors and even of museum staff. In 2014, for preparing the Special Exhibition for the Horse Year, Prof. Deng T. (IVPP, famous paleontologist) was invited to be the advisor to the exhibition and gave lecture on the origin and evolution of horse, which highly promoted the academic level for preparing the exhibition. At the same time, British FRS Edwards D. and Belgian famous expert Prof. Godefroit P. were also invited to attend the exhibition.

2.5.3 Popularization conference and training

Entrusted by the Committee of Scientific Popularization of the Palaeontological Society of China (PSC), the PMOL undertook several popularization conferences and training classes for relevant professional workers and leaders of the museums in China, such as The First Conference of Scientific Popularization in Geology & Palaeontology in China (2011); the first and second Professional Training Class of Curators of Geological & Palaeontological Museums in China in 2013 and 2015, respectively. In the professional training classes

14–16 senior experts, including academicians of China, e.g. professors Yin H. F., Liu J. Q., Shu D. G. and Zhou Z. H., were invited to give lectures, which played positive role in promoting the professional level and scientific popularization in the natural museums of China, and received favorable comments from the participants.

2.6 Communication and cooperation

In response to the trend of internationalization in the palaeontological and geological museums, the PMOL has put a high value both on international and domestic exchanges and cooperation. Up to now, the museum has received hundreds of experts and students from more than 20 countries, including the US, Germany, Russia, UK, France, Belgium, Austria, Hungary, Australia, Japan, DPRK, ROK, Mongolia, Vietnam, Thailand, Canada, India, Afghanistan, Pakistan, Brazil, and Israel. These activities aimed at making academic and museum exchange, hosting large and important symposia, developing co-projects and jointly publishing cooperative publications. To support the PMOL development, Prof. Dilcher D. L. (NAS, US) donated over 6,000 professional books and 30,000 reprints of his own to the PMOL, for which the museum established the Dilcher Library; Prof. Ashraf A. R. (Germany) donated over 3,000 professional books and microscopes of his own to the museum, for which the PMOL set up the Ashraf Laboratory.

Domestically, SNU (including the PMOL) signed agreements on Innovative Cooperation of Palaeontological Study in Northeast Asia with Nanjing Institute of Geology and Palaeontology, CAS (NIGPAS) and Institute of Vertebrate Paleontology and Paleoanthropology, CAS (IVPP); and highly developed academic exchanges with Beijing Natural History Museum, Nanjing Palaeontological Museum, Paleozoological Museum of China, Chongqing Natural Museum, Zigong Dinosaur Museum, Shenzhen Palaeontological Museum, Xinjiang Geological Museum, Geological Museum of Heilongjiang, Jiaying Dinosaur Museum of Heilongjiang, Geological Museum of Jilin University, Hebei Geological University Museum, etc. All these exchanges and cooperation have helped the institutions and museums develop in communication, research and popularization.

2.6.1 International cooperation

For five years, cooperative projects on the studies of the Mesozoic and Cenozoic paleontology and geology, by the PMOL and institutions and museums from Germany, Russia, South Korea, the United States and Belgium, have achieved important results. The project Miocene flora of Changbai in East Jilin and comparison with Europe (CZ654 of Sino-German Science Center, 2011–2013) firstly refreshed knowledge of the Badaogou flora about the Pliocene in age, which greatly stimulated the study of the Cenozoic biostratigraphy in

NE China, and promoted the joint exhibition of the fossils in Liaoning in collaboration with Stuttgart Museum of Natural History, Germany. The Sino-German co-project on the joint investigation on Mesozoic startigraphy in Xinjiang (funded by NSFC, 2014) also made great achievement. Moreover, the PMOL led international cooperative project Late Cretaceous-Paleocene biotas and the K-Pg boundary in Heilongjiang, China (funded by NSFC et al., 2002–2011), and first made the definition of the non-marine K-Pg boundary in China, joined by foreign experts from Russia, Germany, US, UK, Belgium, Japan and Brazil. The co-study of early angiosperms has been in progress by the PMOL and US institutions as well.

In addition, international academic conferences have been successfully organized by the PMOL during the past five years, represented by the Int'l Symposium of Geology & Paleontology in Yichun (Aug., 2011; attended by over 100 experts from 15 countries); the 12th International Symposium on Mesozoic Terrestrial Ecosystems (MTE-12; Aug., 2015 in Shenyang; attended by over 160 experts from 18 countries), which all left deep impressions on the participants for high quality scientifically and the organizations. Besides having academic exchange programmes with its domestic counterparts, the museum has also sent more than 20 staff to Germany, Japan, Austria, Israel, Russia, and US to attend international academic conference in exchange programmes. Prof. Sun G., Director of the museum was invited to the Far-East Federal University of Russia to give lecture in Sept., 2014.

2.6.2 International exhibitions

To develop international cooperation, exchange and scientific popularization on palaeontology abroad, the PMOL has co-organized the Special Exhibition of Feathered Dinosaurs in Liaoning, China with the Orleans- and Nantes Natural History Museums of France, and the Stuttgart Natural History Museum of Germany, during 2012–2015. The exhibitions were held successfully, attracting thousands of European people, which not only popularized the scientific achievements of China in paleontological study, but also greatly increased cultural exchange and understanding between Chinese and European people.

Furthermore, the establishment of the PMOL can be regarded as a "bridge" linking international museums with institutions. A telling example was that in the late 2011, the Japanese delegation of Josai University led by Chancellor Prof. Mizuta N., visited the PMOL and was left with a deep impression of the museum's role in research and education. After her visit, she came up with a plan to build a museum in the Josai University, named the Mizuta Memorial Museum, mainly for the exhibition of fossils. With the help of Prof. Oishi M. (Tokyo Univ., the son of famous paleobotanist Prof. Oishi S.), and Prof.

Sun G., director of PMOL, she was very successful in preparing the building of the museum. The Mizuta Memorial Museum of Josai University was completed and opened in Apr., 2013. Prof. Sun and the PMOL delegation were invited to attend the opening ceremony, which promoted the cooperation between SNU and Josai University.

2.6.3 Domestic exchange and cooperation

The PMOL has paid much attention to the domestic exchange and cooperation. For the past five years, the biggest task carried out successfully for the PMOL has been hosting the 28th Annual Conference of PSC (in Shenyang in Aug., 2015) co-organized with the NIGPAS. Present at conference were more than 450 experts of 90 institutions from 26 provinces and the participants submitted over 300 papers, and 46 posters. Six plenary lectures were given by senior experts, and 215 oral reports were also very significant. In this conference the PSC issued the 3rd Youth Palaeontologists Prize, and the participants visited the PMOL, plus field trips to Beipiao, Chaoyang and Benxi, respectively.

It should be mentioned that in July, 2013, the PMOL hosted successfully the Symposium on Front Study of Jehol Biota, attended by about 100 experts of 23 institutions in collaboration with the Palaeontological Societies of Jiangsu, Henan, Anhui and Hebei provinces, and Jiangsu Geological Society.

2.7 Personnel training

To better develop the museum, SNU and the PMOL have attached great importance to the personnel training programmes of the museum. Up to now, the researchers include 5 professors, 4 associated professors, 5 lecturers and 2 research assistants, of whom 16 have received Ph.D. degree. Besides, 21 technicians work with the researchers, most of them having Master degree, and the majority being young people.

Among the researchers, the leading scientist Prof. Dr. Sun G. is the Corresponding Member of Botanical Society of America (BSA, US), and honored as the National Excellent Scientist of China; Prof. Dr. Zhou C. F. is the winner of Outstanding Young Talent of MLRC; Prof. Dr. Hu D. Y. is honored Excellent Expert of Liaoning Province; and two young researchers are selected into the Talent Training Programme of Liaoning Province.

Besides, there are 20 honorary and guest professors in total in the museum, including six academicians: Dilcher D. L. (NAS, US), Edwards D. (FRS, UK), Akhmetiev M. (Corr. Mem., RAS, Russia), Chinese academicians Li T. D., Liu J. Q. and Zhou Z. H. (President of IPC). The participation of the above experts has played important role in offering consultancy service and supporting the research, professional training and popularization of the PMOL.

To highlight the working experience in the personnel training programmes of the museum, we emphasize the following aspects.

(1) Introduction of excellent academic leaders

With the joint efforts by SNU and DLRL, Prof. Dr. Sun G., the famous Chinese palaeontologist, was introduced from Jilin University (JU) to SNU in 2008, as the director of PMOL. He is not only an excellent scientist but also an experienced administrator, being active on internationally academic stage. The introduction of Prof. Sun laid a solid foundation for the construction and operation of the PMOL.

(2) Development of young professionals

For recent years, the PMOL has successively engaged a batch of excellent young and midaged experts from home and abroad, such as Hu D. Y. (Ph.D. in Hokkaido Univ., Japan), Dr. Duan Y. (Ph.D. in Deakin Univ., Australia), Zhou C. F. (Ph.D. in Peking Univ.), Prof. Dr. Cao C. R. (Ph.D. in JU), Drs Zhang Y., Tian N. and Zhao X. (NIGPAS), and Dr. Liu Y. S. (CNU), etc. The participation of the younger scientists have injected vitality into the PMOL.

(3) Bringing out full potentials of guest professors

The museum pays much attention to the advice and supports of guest professors. All the 20 honorary and guest professors made great contributions to the research, talents cultivation and international cooperation of the museum. Special mention should be made of professors Dong Z. M., Xu X. and Hou L. H. for their advice on the study of dinosaurs and the origin of birds; of Prof. Zhang L. J. on geology of Liaoning. American Prof. Dilcher D. L. (NAS) has devoted to the cooperative study of early angiosperms; Russian Prof. Akhmetiev M. advised on the project study of the K-Pg boundary in Heilongjiang, China, and donated to the museum exhibition precious fossil specimen which was collected by himself in the Moscow River side when he was 11 years old.

(4) Training of technicians

The training of technicians has also been given great attention by the PMOL. The 21 technicians, respectively belonging to seven technical departments, engage themselves in fossil excavation and repair, laboratory, exhibition designing, scientific popularization, collection management, SEM & TEM lab, and informative technology. Most of the technicians have had the opportunity to further their study at home or abroad, such as participating fossil exhibitions aboard, attending training classes in overseas or domestic museums. All the professional trainings not only improve their skills but also enrich their knowledge on geology, palaeontology and museology.

2.8 Laboratories and museum management

2.8.1 "Hardware" construction

There are ten laboratories in the PMOL, including the labs of Paleobotany, Palynology, Micropaleontology (2 in number), Grinding, Fossil Repair, SEM & TEM, Geology, Designing, and Printing.

The SEM & TEM Lab has advanced electron microscopes, including SEM (S-4800), and TEM (H-7700), both made in Japan. Among the optical microscopes, the advanced super-depth 3D microscope (VHX-600E) can directly observe the microcosmic features of the fossils, including structural cells, without chemical macerations. Between 2011–2013, the microscope VHX-600E played a unique role in the study of the epidermal structure of the fossil leaves of *Yanliaoa* from the Middle Jurassic, and an achievement made by Tan X., a graduate of the PMOL, won an international prize in Germany in 2013.

2.8.2 Platform construction

With the support and approval from DLRL and SNU, the PMOL established the Key Laboratory of Evolution of Past Life & Paleoenvironmental Changes in Liaoning Province in 2012, and Key Laboratory of Evolution of Past Life in Northeast Asia, MLRC in 2015. Meanwhile, supported by the museum, SNU first founded the College of Paleontology, assisting the PMOL in training students and research (see 4.1).

2.8.3 Museum management

The PMOL has been affiliated to the DLRL and SNU under the Management Committee of PMOL. The composition of the Committee is as follows:

Director: Ji F. L. (Director of DLRL and Honorary Director of PMOL)

Vice Director: Lin Q. (President of SNU)

Members：Ma Y. (Vice-Director of DLRL)

　　　　　Wang D. C. (Vice-President of SNU)

　　　　　Sun G. (Director of PMOL)

　　　　　Guo J. (Director of BFPL)

　　　　　Yang J. (Director of SNU General Office)

To better run the management of the museum, the PMOL has strengthened the construction of the administrative systems. Over the past five years, to meet museum work needs, dozens of working systems have been formulated, aiming at work responsibility, performance review, financial management, laboratory management and collection management. These mainly include 12 items of work responsibilities of PMOL, and 33 items of management system of PMOL.

Chapter 3
Fossil Protection in Liaoning

3.1 Fossil protection in Liaoning

Liaoning province enjoys an international reputation for the most precious fossils in the world, such as those in the Jehol Boita and Yanliao Biota, as well as numerous results of scientific research in high level world-wide. As a result, Liaoning becomes one of the most important fossil protection sites both domestically and internationally. In 2013, Chaoyang, Yixian and Jianchang of western Liaoning were officially designated as the National Centralized Important Localities for Fossil Protection in China, by the MLRC.

For the past decades, although people's awareness of the fossil protection has been raised, cases of some illegal mining of fossils have frequently been found in western Liaoning, and driven by economic interests, some lawbreakers make profit from selling fossils. Moreover, unlawful people illegally transport fossils abroad, which causes severe damage to fossil localities.

How to protect fossils and their localities, and make people instinctively understand and observe law? How to reasonably develop the study of fossils and localities? All the questions have become the challenges facing governments and specially a new project to the DLRL, particularly for the Bureau of Fossil Protection, Liaoning (BFPL) which is mainly in charge of fossil protection in Liaoning Province.

The fossils protection in Liaoning Province is the earliest programme implemented in China. For 20 years, the governments at all levels in Liaoning have made great achievements and gathered experience through unremitting efforts in protecting fossils. In 1997, Liaoning firstly established fossil protection zone in China. In 2001, it was the first to promulgate Regulation on Protection of Fossils in Liaoning, draw up Grading Standard on Fossils and to build up the Appraisal Council of Fossils of Liaoning. A lot of fossil protection work carried out in Liaoning laid a solid foundation for promulgating Regulation on

Protection of Fossils in China and measures for implementation. Therefore, Liaoning was honored as "model province in fossil protection" in China.

The main working experience gathered by Liaoning in protecting fossils could be summarized from the following aspects: (1) Carrying out fossils and locality investigation, by the DLRL; identifying 247 fossil localities in provincial fossil resource investigation, in which 165 sites yielding important fossils; drawing the Fossil Resource Map in Liaoning Province; and conducting 1∶100,000 geological survey related to fossils, and making more specific geological mapping in key areas for fossil protection. (2) Establishing protection areas and geological parks to reinforce fossil protection, such as first establishing the Beipiao Natural Protection Area of Fossil Birds (1997), the Protection areas or geological parks in Yixian (2002), Chaoyang (2003), Benxi (2006). In 2004, approved by the MLRC, the Chaoyang National Geological Park of Fossil Birds was established, and the protection areas or geological parks were established in Jianchang and Yixian, in 2008, 2010, respectively. (3) Promulgating laws and regulations. (4) Closely following experts' guidance, which is mainly evidenced by the first establishment of the Appraisal Council of Fossils of Liaoning (2001), and the Liaoning Provincial Expert Committee of Fossil Protection (2014), which is composed of over 30 China's senior paleontologists, and some fossil management experts, who bring in new energy, advice, and help for the fossil protection in Liaoning. (5) Furthering museum development which makes for the fossil protection. By Dec., 2015, 12 geological or paleontological museums had already been built with 4 in construction in Liaoning. Therefore Liaoning has become the province with the most geological or paleontology museums in China.

3.2 Co-building the PMOL

To follow the instructions by the central government in fossil protection and creatively complete the fossil protection work in Liaoning, since 2005, the DLRL and SNU have jointly built the PMOL, making great efforts in various aspects, such as policy support, financing, construction guide, etc. On May, 2011, a splendid paleontological museum achieving the first level both at home and abroad was built and unveiled. Mr. Xu S. S., Director of National Commission of Development and Reform, China, and Ex-Minister of the MLRC, commented that the establishment of the PMOL by the DLRL and SNU was a successful model of co-founding museums by government and university (see 2.2).

Chapter 4
College of Paleontology built in Liaoning

4.1 New cradle for training young paleontologists

The major of paleontology in universities of China was established almost a hundred years ago. It was first inserted into the curriculum by famous paleontologist Prof. Li S. G. (J. S. Lee) in Peking University (PU) in 1920s and early 1920s after his return from University of Birmingham, UK, with the assistance of American geologist Prof. Grabau A. W. teaching in the PU at that time. Before 1949, most of Chinese academicians in paleontology had been graduates from the PU. In the 1950s, the major of paleontology was taken in some universities with geological discipline, such as the PU, Nanjing Univ., Beijing College of Geology (now China Univ. Geosci.), Changchun College of Geology (now JU), etc. However, from the late 1960s to 1970s, due to the interferences of the Cultural Revolution, as well as the recession of geological industry, paleontology major was canceled in all above-mentioned universities. Hence the graduates in paleontological major had various background in the undergraduate stage, which posed some problems for the further cultivation of the talents in paleontology. However, in recent years, with the growing need for paleontological experts in the research of petroleum, geology, museum and other areas, it is urgent to reinstate or establish the major of paleontology in universities that will be capable of cultivating experts in this major.

For this reason, Prof. Sun G., Directors of PMOL and Paleontological Institute of SNU, proposed to establish College of Paleontology in SNU, which won support from the SNU leaders. In Dec., 2010, the new College of Paleontology (CP-SNU) was officially established, and it was the only college of paleontology in China, even in the world. In 2011, Sun G. was appointed as the Dean of the College by SNU, and the famous paleobotanist Dilcher D. L. (NAS, US) as the invited Honorable Dean. On Oct. 9, 2011, the grand inaugural ceremony of the CP-SNU was held in Shenyang, and SNU had the first batch of undergraduates in paleontological major who bloomed together with the PMOL.

After five years' development, the college has emerged as a new force in undergraduate education and the cultivation of professionals. The college offers three choices in academic research, namely Paleontology, Fossil Protection & Museology, and Fossil Fuels Research, enrols about 30 undergraduates and 5 graduates every year; and now 113 undergraduates and 15 graduates are currently pursuing their study. The college has a team of high-level teachers, including 5 professors, 4 associated professors, and 5 lecturers, of whom 16 have obtained Ph.D. degree.

Up to now, the college has made more achievements in scientific research, teaching, international exchange and cooperation. The faculty have published 6 papers on *Nature* (3 of them as the first author), and had more than 50 papers indexed by SCI and 8 monographs and textbooks. The research on the early angiosperms of western Liaoning has won the Awarded of Science & Technology, Liaoning (First Prize), and the faculty have received 8 projects granted by the NSFC (China) and other financial resources at ministerial-provocial level. In teaching, the college highlights features in practice concerning paleontology and geology, such as the skills of cuticular and palynological preparations of fossil plants, slicing rocks and wood fossils, SEM technique, etc., all of which have laid a solid foundation for students in research and applications. Besides, in summer semesters teaching practice, students have made geological field trips to Benxi (Liaoning), Qinhuangdao (Hebei), Jiayin (Heilongjiang), Junggar and Turpan (Xinjiang), even to Far Eastern Russia for international geological field practice, which greatly opened students' eyes and improved their ability in application and international communication.

Glorious flowers in spring and solid fruits in autumn. In the past five years, students from two classes in three grades have graduated from the college. A total of 23 students enrolled in 2011 and 2012 have met the graduation qualifications (ca. 43% of all graduates), including of Univ. Glasgow (UK), the NIGPAS, IVPP, PU, China Univ. Geosci. (Beijing & Wuhan), and other institutions; 4 graduates have been candidates for Ph.D. study in IVPP, NECU, and JU. Graduate Tan X. has won a prize in Sino-German Paleontological Conference held in Goettingen, Germany in 2013; undergraduate Wang L. published a paper in *Global Geology* in 2014, and won the first prize of the Award of Innovation Projects of Liaoning Univeristies. In recognition of their professional knowledge and talent in the field of natural museums and cultural industry, 18 students enrolled in 2011 and 2012 have been engaged to work for Chongqing Natural Museum, Anhui Geological Museum, Changzhou Museum of Dinosaur Park, Xinghai Paleontological Museum of Dalian, Shanghai FiveM Animation Studios (Ruihong), Cult. Creat. Ind. Inn. Base of Beijing

Tiantu, and Changzhou Zhuojin Culture Media Company, etc.

4.2 The college with PMOL

The CP-SNU and PMOL have been compared to "twins" for their close cooperation in paleontological research, popularization and training, and they both have witnessed rapid development of education and museum construction in China. Students (including graduates) and teachers of the CP-SNU have provided strong support for the PMOL in scientific popularization and training. All the students and graduates have acted as volunteers of the museum, and played a vital role in popularization. Teachers have been engaged as advisors and commentators in the training classes of the museum, etc. Dr. Li L., paleoavian expert, chaired lectures in Class of Juvenile Interpreters; Dr. Tian N., the young paleobotanist, actively joined in scientific popularization in the countryside in Zhangwu County of Liaoning. All the assistances have received very warm welcome from the audience, especially from teenagers. Other than scientific popularization, the teachers of the CP-SNU have also developed Thursday Class as professional learning programme for museum staff, especially for interpreters and technicians. The close cooperation between the college and museum, on the one hand, has facilitated museum development by using teaching and research resources of the university and on the other hand, turned the museum into a platform and "window" for all the SNU teachers and students to carry out their scientific popularization and related work.

In summary, the Liaoning Palaeontology Museum (PMOL), as a five-year-old "baby" in the big family of national natural museums, is like a rising star. With care and support of scientists and people from all circles of societies, we hope that the PMOL can add its special glory to the development of paleontology and natural museums in the international arena.

参考文献/Reference

[1] 安太庠,张放,向维达,等.1983.华北及邻区牙形石.北京:科学出版社,1—223
[2] 陈芬,孟祥营,任守勤,等.1988.辽宁阜新和铁法盆地早白垩世植物群及含煤地层.北京:地质出版社,1—180
[3] 陈均远,周桂琴,朱茂炎,等.1996.澄江动物群.台中:国立自然科学博物馆,1—222
[4] 陈孟莪,肖宗正.1991.峡东地区上震旦统陡山沱组发现宏体化石.地质科学,4:317—324
[5] 段吉业,曹成润,段冶,等.2015.华北板块东部早古生代动物群,沉积相及地层多重划分.北京:科学出版社,1—401
[6] 郭鸿俊,段吉业.1978.冀东北及辽西寒武纪及早奥陶世新三叶虫.古生物学报,17(4):445—458
[7] 郭鸿俊,段吉业,安素兰.1982.中国华北地台区寒武系与奥陶系的分界并简述有关三叶虫.长春地质学院学报,3:9—28
[8] 顾知微.1962.中国的侏罗系和白垩系.北京:科学出版社,1—84
[9] 洪友崇.1983.北方中侏罗世昆虫化石.北京:地质出版社,1—223
[10] 洪友崇.1998.中国北方昆虫群的建立与演化序列.地质学报,72(1):1—10
[11] 洪友崇,阳自强,王士涛,等.1980.辽宁抚顺煤田地层及其古生物群研究.北京:地质出版社,1—98
[12] 侯连海.1995.中国中生代鸟类.南投:凤凰谷鸟园,1—228
[13] 侯先光,伯格斯琼,王海峰,等.1999.澄江动物群.昆明:云南科技出版社,1—170
[14] 华洪,张录易,张予福,等.2001.高家山生物群化石组合面貌及其特征.地层学杂志,25(1):13—17
[15] 黄力强.2013.庙后山——东北第一人的故乡.沈阳:辽宁民族出版社,1—144
[16] 季强,姬书安.1996.中国最早鸟类化石的发现及鸟类的起源.中国地质,(10):30—33
[17] 郎嘉彬,王成源.2007.本溪组命名剖面牙形刺动物群的特征及地质时代.世界地质.26(2):137—145
[18] 李廷栋.1982.亚洲地质.北京:地质出版社
[19] 李星学.1963.中国晚古生代陆相地层.北京:科学出版社,1—168
[20] 辽宁省博物馆,本溪市博物馆.1986.庙后山——辽宁本溪市旧石器文化遗址.北京:北京文物出版社,1—102
[21] 林启彬.1976.辽西侏罗系的昆虫化石.古生物学报,15(1):97—116
[22] 林英锓,武世忠,邱翠珍.1992.异珊瑚研究的新进展.古生物学报,31(4):488—500
[23] 刘发.1987.辽宁本溪地区本溪组下部腕足类化石的发现及其意义.长春地质学院学报,

17(2):121—154
[24] 刘宪亭,马凤珍,王五力.1987.辽宁西部晚中生代鱼化石.北京:地质出版社,223—234
[25] 卢衍豪,张文堂,朱兆玲,等.1965.中国的三叶虫(上,下).北京:地质出版社,1—362
[26] 孟祥营,陈芬,邓胜徽.1988.杉木属的一个化石种——亚洲杉木.植物学报,30(6):649—654
[27] 米家榕,孙克勤,金建华.1990.辽宁本溪早石炭世植物化石.长春地质学院学报,4:361—368
[28] 米家榕,张川波,孙春林,等.1993.中国环太平洋带北段晚三叠世地层古生物及古地理.北京:科学技术出版社,1—219
[29] 南润善,常绍泉.1982.辽南早寒武世石桥组三叶虫.中国地质科学院沈阳地质矿产研究所所刊,4:7—15
[30] 蒲荣干,吴洪章.1985.辽宁西部中生代孢粉组合及其地层意义.见:张立君,蒲荣干,吴洪章(著).辽宁西部中生代地层古生物(2).北京:地质出版社,121—212
[31] 蒲荣干,吴洪章.1995.东北地区中生代孢粉植物群及其地理分区.见:王五力,郑少林,张立君,等(编).中国东北环太平洋带构造地层学.北京:地质出版社,221—236
[32] 任东,高克勤.2002.内蒙古宁城道虎沟地区侏罗纪地层划分及时代探讨.地质通报,21(8):584—591
[33] 任东,卢立伍,郭子光.1995.北京与邻区侏罗—白垩纪动物群及其地层.北京:地震出版社,1—222
[34] 斯行健,李星学,等.1963.中国中生代植物.北京:科学出版社,125—126
[35] 孙春林,王丽霞,张立军,等.2009.辽宁北票羊草沟上三叠统植物化石新材料.中国古生物学会第25届学术年会论文摘要集,162
[36] 孙革,曹正尧,李浩敏,等.1995.白垩纪植物群.见:李星学(主编).中国地质时期植物群.广州:广东科技出版社,310—341
[37] 孙革.1993.中国吉林天桥岭晚三叠世植物群.长春:吉林科技出版社,1—157
[38] 孙革,孟繁松,钱立君.1995.三叠纪植物群.见:李星学(主编).中国地质时期植物群.广州:广东科技出版社,229—259
[39] 孙革,张立君,周长付,等.2011.30亿年来的辽宁古生物.上海:上海科技教育出版社,1—176
[40] 孙永山.2009.辽西热河生物群古生物化石地质遗迹的基本情况和保护现状.化石,3:32—38
[41] 田宁,王永栋,张武,等.2014.辽西侏罗纪紫萁根茎化石新材料(*Ashicaulis wangii* sp. nov.)及古生物地理学和演化意义.中国科学:地球科学,44(10):2262—2273
[42] 田宁,谢奥伟,王永栋,等.2015.辽西建昌地区侏罗纪髫髻山组木化石的发现及古气候意义.中国古生物学会第28届学术年会论文摘要集,143
[43] 王五力.1987.论中国北方早白垩世早期阜新生物群.中国地质科学院沈阳地质矿产研究所文集,16:53—59
[44] 王五力,郑少林,张立君,等.1995.中国东北环太平洋带构造地层学.北京:地质出版社
[45] 王五力,张宏,张立君,等.2004.土城子阶、义县阶标准地层剖面及地层古生物、构造—火山作用.北京:地质出版社,1—514
[46] 徐仁.1976.铁细菌和富铁矿.化石,(1):5—7
[47] 杨建杰.2015.生命从远古走来——探秘辽宁史前世界.沈阳:辽宁科技出版社,1—112
[48] 杨建杰.2015.生命从远古走来——龙鸟传奇.沈阳:辽宁科技出版社,1—112

[49] 尹磊明.1977.辽东鞍山群、辽河群的微体植物群及其地层意义.中国科学院铁矿地质学术会议论文选集.地层古生物.北京:科学出版社,39—60

[50] 张殿双.2000.探讨我国古生物化石资源的保护与管理.地质问题研究,11:74—76

[51] 张俊峰.2002.道虎沟生物群(前热河生物群)的发现及其地质时代.地层学杂志,26(3):173—177,215

[52] 张立君.1985.辽宁西部晚中生代非海相介形类动物群.见:张立君,薄荣干,吴洪章(著).辽宁西部中生代地层古生物(2).北京:地质出版社,1—120

[53] 张立君,张英菊.1982.辽宁阜新盆地阜新组介形虫化石.古生物学报,21(3):362—340

[54] 张立军.2013.省级古生物化石保护规划编制指南主要内容解读.中国古生物学会第27届学术年会论文摘要集,221—222

[55] 张武.1982.辽宁凌源晚三叠世植物化石.中国地质科学院沈阳地质矿产研究所文集.3:187—197

[56] 张武,张志诚,郑少林.1980.植物界.东北地区古生物图册(二)——中新生代分册.北京:地质出版社,221—271

[57] 张武,董国义.1983.东北地区的三叠系.中国地质科学院沈阳地质矿产研究所文集,8:1—56

[58] 张武,郑少林.1983.辽宁本溪中三叠世林家植物群的研究.中国地质科学院沈阳地质矿产研究所文集,8:62—91

[59] 张武,郑少林.1984.辽西金岭寺—羊山盆地上三叠统老虎沟组植物化石新材料.古生物学报,23(4):382—393

[60] 张武,郑少林.1987.辽宁西部地区早中生代植物化石.见:于希汉,王五力,刘宪亭,等(著).辽宁西部中生代地层古生物(3).北京:地质出版社,239—338

[61] 张武,李勇,郑少林,等.2006.中国木化石.北京:中国林业出版社,1—356

[62] 赵传本,叶得全,魏德恩,等.1994.中国油气区第三系(Ⅲ)——东北油气区分册.北京:石油工业出版社

[63] 赵毅宾,张立军,刘雪飞.2004.辽西热河生物群化石资源的保护管理与可持续利用研究.国土资源科技管理,21(3):33—36

[64] 赵育文.2012.抚顺煤盆地含煤地层及聚煤分析.科技信息,23:287—401

[65] 郑少林.2000.冀北辽西地区土城子组的划分及时代.第三届全国地层会议论文集.北京:地质出版社,227—232

[66] 郑少林,张武.1984.阜新海州组侧羽叶属一新种及其表皮构造.植物学报,26(6):664—667

[67] 郑少林,张武.1990.辽宁田师傅早中侏罗世植物群.辽宁地质,(3):212—237

[68] 中国古生物学会.2015.中国古生物学学科发展史.北京:中国科技出版社,1—328

[69] 周惠琴.1981.辽宁北票羊草沟晚三叠世植物化石组合的发现.中国古生物学会第12届学术年会论文选集.北京:科学出版社,147—152

[70] 朱兆玲.1959.华北及东北崮山统三叶虫动物群.中国科学院南京地质古生物研究所集刊,60(3):373—384

[71] Dilcher D L, Sun G, Ji Q, et al. 2007. An early infructescence *Hyrcantha decussata* (comb. nov.) from the Yixian Formation in northeastern China. *PNAS*, 104(22): 9370−9374

[72] Doyle J A. 2012. Molecular and fossil evidence on the origin of angiosperms. *Annual Review of Earth and Planetary Sciences*, 40: 301−326

[73] Duan S, Wang X, 1997. A peculiar fossil leaf *Beipiaophyllum* Duan & Wang gen. nov. *Chenia*, 3-4: 125-131

[74] Hu D Y, Hou L H, Zhang L J. 2009. A pre-*Archaeopteryx* troodontid theropod from China with long feathers on the metatarsus. *Nature*, 461: 640-643

[75] Hu D Y, Xu X, Hou L H. et al. 2012. A new enantiornithine bird from the Lower Cretaceous of western Liaoning, China, and its implications for early avian evolution. *Journal of Vertebrate Paleontology*, 32(3): 639-645

[76] Li C W, Chen J Y, Hua T. 1998. Precambrian sponges with cellular structures. *Science*, 279: 879-882

[77] Li L, Wang J Q, Zhang X. et al. 2012. A new enantiornithine bird from the Lower Cretaceous Jiufotang Formation in Jinzhou Area, Western Liaoning Province, China. *Acta Geologica Sinica (English edition)*, 86(5): 801-840

[78] Liu Y S, Shin C K, Ren D. 2011. A new lacewing (Insecta: Neuroptera: Grammolingiidae) from the Middle Jurassic of Inner Mongolia, China. *Zootaxa*, 2897: 51-56

[79] Luo Z X, Yuan C X, Meng Q J, et al. 2011. A Jurassic eutherian mammal and divergence of marsupials and placentals. *Nature*, 476: 442-445

[80] Quan C, Liu Y, Utescher S. 2011. Paleogene evolution of precipitation in northeastern China supporting the middle Eocene intensification of the East Asian monsoon. *Palaios*, 26: 743-753

[81] Ren D. 1998. Flower-associated Brachycera flies as fossil evidences for Jurassic angiosperm origins. *Science*, 280: 85-88

[82] Su K., Quan C, Liu Y S. 2014. *Cycas fushunensis* sp. nov. (Cycadaceae) from the Eocene of Northeast China. *Review of Palaeobotany and Palynology* 204: 43-49

[83] Sun C L, Li T, Na Y L, et al. 2014. *Flabellariopteris*, a new aquatic fern leaf from the Late Triassic of western Liaoning, China. *Chinese Science Bulletin*, 59(20): 2410-2418

[84] Sun G, Akhmetiev M, Markevich V, et al. 2011. Late Cretaceous biota and the Cretaceous-Paleogene (K-Pg) boundary in Jiayin of Heilongjiang, China. *Global Geology*, 14(3): 115-143

[85] Sun G, Dilcher D L, Zheng S L, et al., 1998. In search of the first flower: a Jurassic angiosperm, *Archaefructus*, Northeast China. *Science*, 282(5394): 1692-1695

[86] Sun G, Ji Q, Dilcher D L, et al. 2002. Archaefructaceae, a new basal angiosperm family. *Science*. 296(5569): 899-904

[87] Sun G, Dilcher D L, Wang H S, et al. 2011. A eudicot from the Early Cretaceous of China. *Nature*, 471: 682-685

[88] Tian N, Wang Y D, Zhang W, et al. 2013. *Ashicaulis beipiaoensis* sp. nov., a new species of osmundaceous fern from the Middle Jurassic of Liaoning Province, northeastern China. *International Journal of Plant Sciences*, 174(3): 328-339

[89] Tian N, Wang Y D, Zhang W, et al. 2014. A new structurally preserved fern rhizome of Osmundaceae (Filicales) *Ashicaulis wangii* sp. nov. from the Jurassic of western Liaoning and its significances for palaeobiogeography and evolution. *Science China (Earth Sciences)*, 57(4): 671-681

[90] Wang X L, Kellner A W A, Zhou Z H, et al. 2008. Discovery of a rare arboreal forest dwelling flying reptile (Pterosauria, Pterodactyloidea) from China. *PNAS*, 105(6): 1983-1987

[91] Xu X, Zhou Z H, Wang X L, et al. 2003. Four-winged dinosaurs from China. *Nature*, 421: 335-340

[92] Yuan X L, Chen Z, Xiao S H, et al. 2011. An early Ediacaran assemblage of macroscopic and morphologically differentiated eukaryotes. *Nature*. 470: 390-393

[93] Zhang J F. 2006. New mayfly nymphs from the Jurassic in northern and northeastern China (Insecta: Ephemeroptera). *Paleontological Journal*, 40(5): 553-559

[94] Zhang J F, Kluge N J. 2007. Jurassic Larvae of Mayflies (Ephemeroptera) the Daohugou Formation in Inner Mongolia, China. *Oriental Insects*, 41: 351-366

[95] Zhang J F. 2010. Revision and description of water boatmen from the Middle-Upper Jurassic of northern and northeastern China (Insecta: Hemiptera: Heteroptera: Corixidae). *Paleontological Journal*, 44(5): 515-525

[96] Zhang J F. 2015. Archisargoid flies (Diptera, Brachycera, Archisargidae and Kovale-visargidae) from the Jurassic Daohugou biota of China, and the related biostratigraphic correlation and geological age. *Journal of Systematic Palaeontology*, 13(10): 857-881

[97] Zhang J F, Kluge N J. 2007. Jurassic larvae of mayflies (Ephemeroptera) from the Daohugou Formation in Inner Mongolia, China. *Oriental Insects*, 41: 351-366

[98] Zheng S L, Zhang L J, Gong E P. 2003. A Discovery of *Anomozamites* with reproductive organs. *Acta Botanica Sinica*, 45(6): 667-672

[99] Zheng S L, Zhou Z Y. 2004. A new Mesozoic *Ginkgo* from Western Liaoning, China and its evolutionary significance. *Review of Calaeobotany & Palynology*, 131: 91-103

[100] Zhou C F, Wu S Y, Thomas M, et al. 2013. A Jurassic mammaliaform and the earliest mammalian evolutionary adaptations. *Nature*, 500: 163-167

[101] Zhou C F, Rabi M. 2015. A sinemydid turtle from the Jehol Biota provides insights into the basal divergence of crown turtles. *Scientific Reports*, 5:16299

[102] Zhou Z H, Barrett P M, Hllton J. 2003. An exceptionally preserved Lower Cretaceous ecosystem. *Nature*, 421: 807-814

[103] Peng S C, Babcock L E. 2008. Cambrian Period. *In*: Ogg J G, Ogg G, Gradstein F M (eds). *The Concise Geologic Time Scale*. Cambridge: Cambridge University Press, 37-46

[104] 彭善池,侯鸿飞,汪啸风. 2016. 中国的全球层型. 上海:上海科学技术出版社,1—149

[105] 米家榕,徐开志,张川波,等. 1980. 辽宁北票附近中生代地层. 长春地质学院学报,(4):18—37

[106] Huang D Y, Engel M S, Cai C Y, et al. 2012. Diverse transitional giant fleas from the Mesozoic era of China. *Nature*, 483: 201-204

[107] Zhou Z H 2014. The Jehol Biota, an Early Cretaceous terrestrial Lagerstätte: new discoveries and implications. *National Science Review*, 1: 543-559

其他作者简介 The co-authors

胡东宇
古鸟类学专家,辽宁古生物博物馆教授
Prof. Dr. Hu D. Y., Ave Palaeontologist, PMOL

周长付
古脊椎动物学专家,辽宁古生物博物馆教授
Prof. Dr. Zhou C. F., Vertebrate Palaeontologist, PMOL/ SNU

刘玉双
古昆虫学专家,辽宁古生物博物馆副教授
Dr. Liu Y. S., Paleoentomologist, PMOL/SNU

杨 涛
古植物学专家,辽宁古生物博物馆讲师
Dr. Yang T., Paleobotanist, PMOL/SNU

傅仁义
古人类学家,辽宁古生物博物馆教授
Prof. Fu R.Y., Paleoanthropologist, PMOL

程绍利
辽宁省化石局副局长,兼辽宁古生物博物馆副馆长
Mr. Cheng S. L., Vive-Director, BFPL/PMOL

杨建杰
辽宁古生物博物馆书记兼副馆长
Ms.Yang J. J., Vice-Director, PMOL

张洪钢
辽宁古生物博物馆副馆长
Mr. Zhang H. G., Vice-Director, PMOL

孙永山
辽宁省国土厅副巡视员,辽宁省化石局前局长
Mr. Sun Y. S., Vice-Director, DLRL; Ex-Director, BFPL

王丽霞
国家古生物化石专家委员会办公室副主任
Ms. Wang L. X. Vice-Director, General Office, NFEC, China

图书在版编目(CIP)数据

走进辽宁古生物世界/ 孙革等著.—上海：上海科技教育出版社,2016.12
ISBN 978-7-5428-6499-4

Ⅰ.①走... Ⅱ.①孙... Ⅲ.①古生物—介绍—辽宁 Ⅳ.①Q911.723.1

中国版本图书馆CIP数据核字(2016)第246426号

责任编辑　王世平　伍慧玲
书籍设计　汤世梁

走进辽宁古生物世界
孙　革　等著

出版发行	上海世纪出版股份有限公司 上海科技教育出版社 (上海市冠生园路393号　邮政编码200235)
网　　址	www.ewen.co www.sste.com
经　　销	各地新华书店
印　　刷	上海中华印刷有限公司
开　　本	889×1194　1/16
印　　张	13
版　　次	2016年12月第1版
印　　次	2016年12月第1次印刷
书　　号	ISBN 978-7-5428-6499-4/N·986
定　　价	158.00元